TEACHING MATHEMATICS IN PRIMARY SCHOOLS

TEACHING MATHEMATICS IN PRIMARY SCHOOLS

Robyn Jorgensen and Shelley Dole

2ND EDITION

For the *Teaching Mathematics in Primary Schools* website,
go to **www.allenandunwin.com/teachingmaths**

ALLEN&UNWIN

Our thanks to the schools for allowing access to classrooms for photographs.

The authors acknowledge the contribution of Robert J. Wright to the first edition of this book, some of which has been retained or adapted in this new edition.

First published in 2004
This edition published in 2011

Allen & Unwin
83 Alexander Street
Crows Nest NSW 2065
Australia
Phone: (61 2) 8425 0100
Fax: (61 2) 9906 2218
Email: info@allenandunwin.com
Web: www.allenandunwin.com

Cataloguing-in-Publication details are available
from the National Library of Australia
www.trove.nla.gov.au

ISBN 978 1 74175 723 1

Internal design by Bookhouse, Sydney
Cartoons by Dave Kill
Index by Susan Jarvis
Set in 10.75/14 pt Utopia by Bookhouse, Sydney
Printed in China at Everbest Printing Co

10 9 8 7 6 5 4 3 2 1

Contents

About this book

This book has been designed to introduce pre-service and practising teachers to current methods of teaching school mathematics to primary and middle school students. We have drawn on a wide range of research to inform the book but have not taken one specific approach. Rather, our intention has been to expose readers to the diversity of approaches to teaching school mathematics. Our rationale for doing this is that students learn from a range of experiences, and with such diversity in contemporary classrooms, teachers need to have a wide repertoire of skills and knowledge if they are to provide the quality experiences needed for student success. It is recognised that the most important factor in creating success for learners is the teacher. As such, teachers in today's society have a critical role in facilitating successes for all students. By having a rich repertoire, and a strong knowledge based on evidence, teachers are better able to prepare their students for a positive future.

We have opted to use the term 'school mathematics' rather than 'mathematics' because there is increasing recognition that what is taught in schools is a particular form of mathematics. It is quite different from the mathematics used in workplaces or by mathematicians and should be recognised as such. Creating opportunities for students to learn and develop a deep appreciation for mathematical thinking and doing is essential for success in life. School mathematics plays an important role in the preparation of young people for their entry into the adult world.

Trends in research are reflected by the availability or paucity of publications. In this second edition we have sought to update the book with more current literature, and have been surprised by the lack of research in mathematical topic areas. In the first edition, the research field was littered with research around mathematical topic areas such as number, fractions, or geometry. However, in the seven years since the first edition was published, there has been a trend away from larger studies of this nature. Because of this, we have been unable to access significant studies to inform the rewriting of sections. Similarly, we had expected that there would be a large movement towards the integration of ICTs (Information Communication Technology) into the teaching of school mathematics since the first edition. Again, we have been surprised and dismayed by the paucity of research in this area. Where such studies were available, we have made reference to them to update specific sections within the book.

We have spent some time considering the merits of including a special chapter on the use of digital media in the teaching of school mathematics. However, we have opted to include the use of technologies within each chapter rather than as stand-alone chapters. This is so that readers are able to see where and how technologies (digital and others) can be used in the teaching of school mathematics. We would strongly argue that technologies can be useful tools in the enhancement of mathematical learning if used appropriately. They enable deeper explorations of many ideas by taking away the tedium of calculations or drawings but they must be used with care and caution. Balance is the key to success.

A further expansion in this second edition has been the inclusion of the 'middle years', those between upper primary and early secondary. Many schools are now moving to restructure schools to include specific programs for the middle years. For many students, the transition to secondary school was often spent revising much of the mathematics taught in primary school, creating the risk of significant repetition in learning, rather than seeing this transition into the middle years as part of the learning continuum. To help avoid this considerable overlap, we have included the extra school mathematics that is typically taught in these years. Not only then does this edition cater for the clearly defined middle years, but also allows primary school teachers to see where learners will be taken with their mathematical learning. It also allows secondary school teachers to understand where their students may need extra support if they appear to be having difficulties learning new knowledge that is part of the secondary school learning experience. This continuum is important knowledge for quality teachers to recognise if they are to provide quality learning for their students and diversity in their student cohort.

About the authors

Robyn Jorgensen is a Professor of Education at Griffith University where she leads mathematics education teaching and research. Robyn has worked across the various sectors of schooling from early years, primary, middle and secondary as well as vocational and post-compulsory schooling. Her research interests are centred on issues of equity and access particularly for students from disadvantaged backgrounds—low socioeconomic status, rural, remote, and Indigenous students. She has received eight Australian Research Council grants to work in these areas. In 2009–10, she took up a position of CEO and Principal in a remote Aboriginal community. Her work is predominantly based in classrooms where she works to better understand how teaching practices are implicated in the success (and failure) of learners. She has been Chair of the Queensland Studies Authority Curriculum Advisory Board for Mathematics; worked on the Queensland Ministerial Committee for Science, Technology, Engineering and Mathematics (STEM); and is currently the eminent Professor on the Australian Association of Mathematics Teachers' 'Count on Me' project that seeks to improve mathematical learning for Indigenous Australians. Robyn has a distinguished publication record with many books, journal articles and conference papers. She has been invited to be keynote speaker at numerous state, national and international conferences. She is the Editor of the Mathematics Education Research Journal and serves on the editorial board for the *International Journal of Science and Mathematics Education*. While currently living in Queensland, she has also lived in Victoria, New South Wales and the Northern Territory. These experiences have helped to inform her breadth of knowledge of the diversity of Australian society and curriculum.

Shelley Dole is a Senior Lecturer in the School of Education at The University of Queensland, where she coordinates Mathematics Curriculum Studies for prospective primary and middle school teachers in the pre-service Bachelor of Education program. Over the 20 years she has been in education, she has taught in primary, secondary and tertiary teaching institutions in the Northern Territory, Queensland, Tasmania and Victoria. Her research interests include new learning and mathematics curriculum change and innovation; learning difficulties, misconceptions and conceptual change associated with learning mathematics; assessment, intervention and unlearning; rational number topics of ratio and per cent and the development of proportional reasoning and multiplicative structures; mental computation and invented algorithms; and teacher professional development. She has been involved in several major research projects in Queensland, Tasmania, Victoria and South Australia including middle years literacy and numeracy; early years literacy and numeracy and distance education; mental computation, number sense and invented algorithms; teaching and learning per cent in the middle school; basic facts in the early years, as well as teacher professional development projects. She is co-author (together with Alistair McIntosh) of a widely used teacher resource package on mental computation strategies that resulted from a large research project conducted in Tasmania and the Australian Capital Territory (ACT), published by the Tasmanian Education Department. She is currently the Chief Investigator of an Australian Research Council project focusing on numeracy across the curriculum and the development of proportional reasoning, with schools in Queensland and South Australia. In 2009, she won a University of Queensland Award for Teaching Excellence, and in 2010 was the recipient of an Australian Award for University Teaching.

CHAPTER 1

The changing face of school mathematics

There is a need for new approaches to teaching mathematics. Sadly, mathematics is still rated by students as one of their least favourite subjects at school, partly because mathematics teaching is too often typified by students working individually with desks in rows facing the board, and the teacher demonstrating procedures from the textbook with students completing textbook exercises. With the vast amounts of research that have been generated since mathematics education became recognised as a discipline, there is now a strong research base to inform change in school mathematics. Also, wide changes are occurring rapidly in societies—nationally, internationally and globally. Change impacts upon students, and there are new theories of how students learn in contemporary times. Research also has contributed to changing perceptions of mathematics as a discipline.

Mathematics education and society

The mathematics curriculum has not been created in a vacuum. Mathematics in schools, and the way it is taught, are the product of broader factors that extend beyond the classroom. Such factors include employers, lobby groups, government policy, parents and professional organisations.

External authorities, such as education departments and other statutory authorities, may develop a curriculum or syllabus that provides guidelines teachers are expected to use to develop their work programs, and to undertake assessment and reporting of students' learning. The development of curriculum documents is influenced by demands from groups that include parents, employers and governments. In recent times, the Australian federal government has developed an agreement among the states and territories for the development and implementation of a national curriculum, which will be implemented across the nation. Documents relating to the national curriculum can be accessed through the ACARA website (<www.acara.edu.au>). When all variables are considered, the curriculum guidelines that appear in mathematics classrooms have been created through a highly negotiated (and often hotly contested) process.

In some cases, curriculum shaping is a reciprocal process, where the benefits of change are two-way. Consider the impacts of technology within society, and of numeracy expectations. In the current context of schooling, employers are demanding that students exit schools with high levels of numeracy (and literacy). As a result, there is a much heavier emphasis on numeracy in education. Similarly, schools recognise the value of technology as a learning tool so that students will exit schools with a strong appreciation of how technology can be used to enhance working and thinking mathematically.

Teaching mathematics in modern society

We have used the term 'modern society' throughout this book to refer to the contempory context of education. 'Modern' is more appropriate than terms such as 'Western' as it does not support a notion that Western views and approaches are more valid than those of the East or Indigenous cultures. Instead, it suggests that the curriculum reflects a contemporary view of education that embraces new approaches to teaching, such as the use of digital resources (the internet, computers, hand-held devices, and so on).

In the times in which our students live, technology, globalisation, the information age and very different patterns of family, leisure and work have brought changes to society, work, schools and life. We use the term 'modern society' to provoke thinking about the age in which we live and the quite different lives of contemporary young people, and to consider how these have changed from 'old times'. Educational researchers have underscored that modern society is different. Many cultures—such as those that embrace Eastern philosophies, or indigenous cultures seeking to gain access

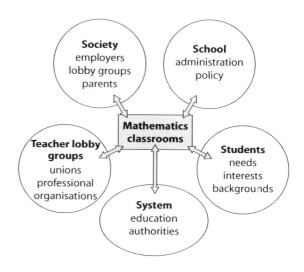

to contemporary ways of thinking and learning—also live in the modern world. The curriculum in schools must reflect the changes occurring in the wider society to ensure that schools adequately prepare students for the world beyond compulsory schooling.

Mathematics classrooms

Most young people are now growing up in technology-rich environments. They do not remember a time when you had to physically get up to change the television station—remote controls do that. Cooking in pre-programmed microwaves happens at the touch of a simple button. They are technologically savvy.

One of the biggest growth areas in employment is self-employment, which means that many young people will be creating jobs for themselves in positions that won't even exist when they exit school. Our students are growing up immersed in an information-rich society—they no longer have to search through the few books in a school or local library, but can undertake searches on the computer that may yield them hundreds of hits. The skills they require in order to be able to search for and identify key information are very different from those they needed when the written word was all that was available.

Students growing up in this technology-rich world have become used to multiple sources of information input—they are constantly bombarded with short bursts of infotainment, as well as brief snippets of information from television and other media. They are able to fragment and reconstruct images (such as maps) in ways not imaginable in old times (Lowrie, 2003). Common terms used to describe contemporary students, such as 'cyberkids', recognise that their dispositions towards learning have been formed by the wider social conditions in which they have grown up. Traditional models of teaching and learning need to reflect these changed circumstances.

The mathematics education of the students of modern society must be considered in light of this. The students need to develop mathematical ways of seeing and interpreting the world; they need to develop strong problem-solving skills; they need to be numerate; and, most importantly, they must have a disposition towards using mathematics to solve the problems they confront. School mathematics needs to adopt pedagogies that will cater for diversity within a classroom. The old models of seated individual work found in traditional mathematics teaching are possibly contributing to the problems that emerge as students progress through school. For considerable numbers of students in the upper years of primary school and lower secondary school, the teaching that they encounter can lead to many negative feelings and misleading learnings about mathematics.

New models of teaching mathematics

The mathematics curriculum encountered prior to the 1960s focused on arithmetic and operations. Most of the mathematics education developed after that time in Western countries emerged post-Sputnik, when the race to the moon had become a race for intellectual superiority, with mathematics seen as the linchpin of success. The 'New Mathematics' contributed to a lock-step approach to teaching mathematics, with hierarchies in orders and sequences of teaching (Brown et al., 1998). The 1970s witnessed a boom in the ways in which mathematics curricula were organised; most were not research-based, but rather influenced by arguments of logic and reason. A hierarchical approach to mathematics teaching ('skill, drill and kill' before application and problem-solving) was implemented in most Western classrooms, and for many teachers and systems such an approach has become a way of life. Brown and colleagues (Brown et al., 1998) argue that much of what was written in terms of mathematics curriculum reform had very little research base to it, thus raising questions about the validity of the curriculum itself.

In more recent times, there has been a growing awareness that such approaches are not resulting in positive learning outcomes. Indeed, as Clements (1989) argues, all students learn from ten years of compulsory schooling (and in most countries this may be extended to twelve to thirteen years) is that they can't do mathematics! Not all countries have bought into this approach to teaching. Research emanating from these countries, particularly the Netherlands (see Anghileri, 2001; Beishuizen, 1999; Buys, 2001; Treffers and Beishuizen, 1999; van den Heuvel-Panhuizen, 2001), has strongly focused on developing new methods and approaches to teaching mathematics.

The Netherlands did not embrace the New Mathematics movement, instead focusing efforts on how students think mathematically (Treffers, 1991). There is now a substantive body of knowledge drawing on students' thinking that has not been constrained by New Mathematics. From this, curricula have been developed that draw on students' understandings, build on them, and progressively move towards abstract and formal mathematical processes. The Dutch mathematics reformers call this 'progressive mathematisation'. Occurring in parallel with this work has been the development of constructivism theory and a general awareness that students actively construct meaning from their experiences. This latter work has had a powerful influence in mathematics education, with it increasingly being recognised that students' individual understandings are based on their lived experiences.

These twin movements have emerged at a time when it is being recognised that many of the old methods of teaching mathematics are failing too many students. This has made the moment ripe for identifying more valid methods of teaching mathematics, and many countries, states and provinces are now adopting new methods of mathematics teaching and learning.

Content and pedagogy

Contemporary approaches to teaching mathematics need to encourage two aspects: content and pedagogy.

Content is the intellectual integrity of the subject. It is where students learn, apply and appreciate mathematics, and where deep learning and deep knowledge are paramount to learning experiences. Importantly, students are able to make connections between the mathematics they learn and other curriculum areas, as well as with the world beyond school. It is important for them to develop an appreciation of how mathematics is an informing discipline that has importance and relevance to many spheres of life.

Pedagogy relates to developing supportive environments where student diversity is recognised and practices are developed that value and build onto the different backgrounds and knowledges that students bring to the mathematics classroom. Good pedagogy is about teachers developing inclusive practices to build and extend their students' knowledge and confidence in using and applying mathematics. Classrooms are places where students understand the expectations teachers have of them and the work they are to undertake. In developing inclusive practices, intellectual integrity also should be preserved. Teachers need to value students and believe that *all* students can learn mathematics.

Research on productive pedagogies (Education Queensland, 2001) states that good pedagogy is about high intellectual engagement and helping students to see and make connections; it is learner-centred, with each individual's knowledge and culture valued, and students feeling supported in their learning. Many teachers bemoan issues of behaviour management in mathematics classes, but in a large longitudinal study on teaching (Education Queensland, 2001) it was found that many of the elements of productive pedagogies were absent from the 2000-plus classrooms observed. Students were not engaged in deep learning about and through mathematics. Often, it was found that students undertook busy work during mathematics lessons (for example, sticking butterflies on paper rather than engaging in discussion about area) but did not engage in much, if any, deep mathematical learning. It must be asked whether students in mathematics classes engage in behaviour that is subsequently construed as 'misbehaviour' because they are bored or because the pedagogy is not sound.

● Teachers can make a difference

Teachers and teaching can make a significant difference to students' learning outcomes in school mathematics. In a large study of effective teachers, Hill and Rowe (1998) reported that it was not the school that made the difference to students, but rather individual teachers. Teachers have a powerful influence over what and how students learn. Through providing the appropriate learning environment in which content and pedagogy match the backgrounds, needs and interests of individual students, *all* students can learn mathematics.

● The power of teachers' beliefs

In a study exploring the characteristics of effective teachers of numeracy, Askew and colleagues (Askew et al., 1997) concluded that one of the most important influences on learning was the teacher's belief that all students could learn mathematics. Often, values and stereotypes influence how behaviours and actions are interpreted and implemented. Teachers who believe that some students, due to their backgrounds or behaviours, are unable to learn mathematics will ultimately create learning environments that construct the expected outcomes. This has been shown to be the case in many studies. For example, in a seminal study (Rosenthal and Jacobsen, 1969), it was shown that when students were assigned scores randomly as they commenced study in a new class at the beginning of term, the teacher, believing that these were the students' academic scores, interacted with different students in particular ways. By the end of the teaching term, the students' results for that class correlated strongly with the scores they had randomly been assigned at the term's beginning.

This study (and many subsequent ones) has highlighted how powerfully teachers' views of their students influence the ways in which they organise learning experiences. In a powerful study in New Zealand, Bishop and Berryman (2006) investigated how to improve the educational achievement of Maori students. If a teacher believes that all students can learn mathematics, then learning environments are likely to reinforce this belief. Similarly, if teachers believe that the best way to learn mathematics is through making strong connections with real-life examples, then the teaching and learning environment will reflect these beliefs. The classroom is likely to be peppered with equipment and displays demonstrating the links between mathematics and the world beyond schools. In good teaching, teachers must be mindful of the influence of their beliefs about teaching and learning mathematics, and the power this has upon the learning outcomes of their students.

What is mathematics?

When a member of the general community is asked this question, they invariably focus on number and operations. However, the modern mathematics curriculum is far more complex than just arithmetic. School mathematics has changed over time. During different periods, different mathematics has been taught. A century ago, mathematics in upper primary classes involved computation of tasks involving a large amount of numbers, long division, square roots of non-square numbers and so on. This form of curriculum remained in place until the early 1960s, when the implementation of the New Mathematics represented a considerable shift in maths curriculum in most Western countries. New topics were included and new forms of thinking mathematically (e.g. set theory) were part of the new syllabuses.

Since the 1970s, other reforms have influenced the curriculum, including problem-solving where students were expected to be more creative in their thinking. In the 1980s and increasingly since, technology has played a role in the curriculum. While the software program LOGO° had a strong influence in the early years of technology-aided learning, it has now been replaced by other technological aids, including new software programs, spreadsheets and graphic calculators, to name a few. Students are expected to be far more creative in their thinking, and to deal with much more knowledge and complexity than in the pre-1960s era.

The interrelationship between society and school mathematics

● Mathematics is the study of patterns and relationships

One of the defining characteristics of mathematics is being aware of recurring ideas and relationships between and among mathematical ideas. The knowledge learnt in one area of mathematics links to other areas—for example, the number facts do not have to be learnt as discrete facts, but rather the student who can see that when you add 3 and 5 there is a total of 8 can also recognise that if you have eight items and three are lost, then there should be five remaining. The place value system can be seen as a patterning of 10 so that by allowing young students to play on a calculator with the constant function set for the addition of 10, they can see the pattern of adding tens occurring in the 10s column.

Seeing these types of patterns and relationships is a key factor in how students come to learn and appreciate mathematics.

● Mathematics is a way of thinking, seeing and organising the world

Students who come to learn mathematics as a dynamic discipline through which much of the world can be interpreted are able to make sense of a wide range of experiences. They are able to organise and analyse events in systematic ways. Sometimes this might be numerically, but it could equally be spatially or through logic. By viewing the world through a mathematical lens, considerable progress can be made in everyday circumstances, such as remembering phone numbers, because a pattern can be seen and people are thus able to solve problems more efficiently and effectively. Mathematics moves beyond memorising a vast body of facts and procedures in a rote manner to become a more systematic and insightful process.

● Mathematics is a language

In past times, mathematics and language were seen as two disparate disciplines. Today it is seen that, in order to learn, appreciate and understand mathematics, students need to learn the language of mathematics—complete with its unique words, grammar and symbols. By knowing the language of mathematics, complex ideas can easily be communicated. Language involves communication, and mathematics is a very particular and precise language that is communicated in particular ways. It involves the contraction of lengthy tasks into short and concise 'sentences'. For example, a shopping

I can remember my friend's phone number because it goes 5522 3612. The first are doubles and then it is 3 and double 3 and double 6 — 3.6.12.

expedition where three items were purchased in one outlet and five in another becomes $3 + 5 = 8$. Abstract symbols such as $+$ and $=$ represent specific concepts.

● Mathematics is a tool

People use mathematics to solve problems every day. The more competent a person is with mathematics, the more efficiently problems can be solved, and in many cases the better one can survive in the world beyond schools. Deciding which goods to purchase, checking bank balances, being able to budget, deciding which mortgage to take out or how much concrete to purchase for the patio all require the use of mathematics. Mathematics enables people to make sound decisions and judgements, and to solve problems.

● Mathematics is a form of art

For most students, mathematics is a grind that has to be done under duress. In contrast, for those who have been fortunate enough to engage with mathematics, its internal beauty is awesome to behold—it has the consistency of a fine artwork. The logical coherence of Fibonacci's numbers, for example, is almost magical. (The Fibonacci sequence is generated by adding the two previous numbers in the sequence to get the next number: 1, 1, 2, 3, 5, 8, 13 . . .). This aspect of mathematics is perhaps the most elusive for many people, since they equate mathematics with the school process of 'doing mathematics' rather than viewing it as something to be appreciated.

● Mathematics is power

Mathematics has been behind most inventions—good and bad—in modern history. It has enabled us to walk on the moon, but also to invent the atomic bomb and mass destruction. Mathematics also gives you the power to take control in your life by ensuring you are given the right change in money transactions, or questioning information presented in mathematical terms, and so on. Those students who succeed in mathematics are more likely to live successful lives than those who don't. Mathematics is the social filter that facilitates the access of some students to professions of high status, wealth and power while excluding others. Developing nations actively seek access to mathematics for their youth, for they know that such knowledge will benefit them in the future. Many Western countries are recognising the problems inherent in increasingly fewer students undertaking formal studies in mathematics, for it is the foundation for so many other forms of powerful knowledge. Computing, science, technology and research all have a heavy reliance on mathematics.

Where does mathematics come from?

While there are a number of theories regarding the generation of mathematics, two key but opposing views permeate the discipline. While there are other views, dealt with in considerable detail by Ernest (1991), it is worth considering the two main views and how they impact both on the ways in which mathematics curricula are constructed, and how the discipline is taught. Just as theories of how humans developed create considerable debate, so too do theories about the origin of mathematics. There are both Platonists and constructionists in mathematics. Some of their opposing notions are listed in the table below.

Opposing mathematical notions in Platonist and constructionist theory

Platonist	Constructionist
Mathematics is pre-existing	Mathematics is a construction of knowledge
The role of the mathematician is to 'discover' facts	Mathematicians create new knowledge in order to explain or expand ideas
Mathematics is about truths and objectivity	Mathematics is about creativity and subjectivity
Teaching involves learning of facts	Teaching involves creating learning environments for students to create their own mathematical understandings
Assessment is objective and tests for factual knowledge; e.g. $\frac{3}{10} + \frac{4}{10} = \frac{7}{10}$ is assessable and true—it is either right or wrong	Assessment is subjective: e.g. $\frac{3}{10} + \frac{4}{10}$ should be $\frac{7}{10}$ but if a student responds with $\frac{7}{20}$, question why. It may be that the student sees the question as being about test results from two exams and not the addition of two fractions

The mathematics curriculum

Teachers developing school-based programs or shared units of work can attest to the debates that develop in seeking to form a coherent learning scheme. The mathematics curriculum is a socially negotiated and constructed document, and as such represents particular viewpoints of various interest groups. The syllabus that is presented by a particular education authority or through a commercial venture is the product of a group of people working towards the development of an agreed document.

The meaning of the term 'curriculum' is highly contested, ranging from the collection of knowledge that will be implemented formally in a classroom through to the planned and unplanned learning of students coming to school. The more widely agreed notion of curriculum is that of knowledge being implemented so that it resonates with the notion of syllabus. This is then supported through teaching and assessment practices.

● Curriculum design

When planning mathematics curricula, departments of education and other peak bodies tend to organise them around a number of key features. In many cases, these are seen primarily as the content areas, but mathematics is far more encompassing than this. Consequently, these documents often include pedagogical approaches to teaching and assessment; principles of equity and diversity within the mathematics classroom and how best to cater for this aspect of mathematics education; and contemporary moves in practice that reflect the dynamic and changing nature of not only the content but also ways of teaching and assessing mathematics. Often the role of various technologies in mathematics is also included.

● Curriculum organisation

Contemporary terminology for organising mathematics sees the curriculum broken down into content area and levels of difficulty. These general principles for organising curricula have filtered through to most contemporary international curriculum documents, and in Australia since the early 1990s, topics within the mathematics curriculum have been organised under five main headings, identified as strands. The five strands of the curriculum are: Number, Space, Measurement, Chance and Data, and Algebra. Curriculum frequently includes reference to *Working Mathematically* to capture the processes and attitudes that the mathematics curriculum seeks to promote. A general overview of topics assigned to these strands is shown in the table.

NUMBER
- Pre-number work around patterning, one-to-one correspondence
- Number work—cardinal numbers, nominal numbers, ordinal numbers, counting, place value
- Whole number operations (addition, subtraction, mutiplication and division)
- Rational numbers—fractions, common, decimals, percentages, mixed
- Estimation, mental computations
- Number sense
- Ratio and proportion

ALGEBRA
- Pre-algebra patterning
- Equality
- Algebra
- Functions

CHANCE AND DATA
- Probability
- Statistics and data—collect, organise, display

WORKING MATHEMATICALLY
- Attitudes and values
- Mathematical inquiry
- Choosing and using mathematics (modelling, applications)
- Communication, representation and language
- Problem-solving
- Connections with other strands, curriculum areas, beyond school

MEASUREMENT
- Length
- Area
- Volume and capacity
- Mass
- Time
- Money
- Temperature

SHAPE
- Location and arrangement—directions, grid work, coordinates (topological)
- Shape—2D shapes, 3D solids, structures, properties, tessellations
- Transformation, symmetry—changes in position, location

A model for organising curriculum content areas

Number

This refers to the study of pure number. It involves the early study of number and patterning. In order to develop good number sense, understanding how patterns are developed is an important part of mathematical thinking. The early years of schooling place considerable emphasis on developing a good sense of number. Formal study of number and operations occupies a considerable amount of space in the timetable. Teaching whole numbers and operations on whole numbers generally precedes work on part numbers. Conservative estimates indicate that teachers can spend up to 80 per cent of their teaching time in mathematics doing number work.

Measurement

This strand seeks to measure objects. There is potential for considerable slippage in this strand, as the concept of measurement relates to other strands as well. For example, angle could as easily rest in the measurement strand as in the space strand. In some documents, money—which is a measurement of value—is placed in the number strand as it strongly links with students' experiences with number, and notions of decimal fractions are easily applied to number. Similarly, chance and data are measured, so in some documents they are located in this strand. These differences show the arbitrariness of curriculum construction.

Chance and data

This is the strand that works from data to make inferences. Probability is often based on data collected and used to predict outcomes. In order to make effective predictions, data are usually collected or inferred mathematically, presented in a format that makes sense of the data so that informed decisions can be made about the predictability of an event occurring. Since data are collected and the chance of an event is predicted, some curriculum documents have chance and data in the measurement strand.

Space

This strand, the study of how space is filled and described, typically includes shape, geometry and location. This element of the curriculum tends to be a little more fragmented than the other strands, since the sub-strands are not so connected and hence are less open to a coherent framework for development. A considerable body of research documents gender differences in the approach to mathematics, and it is in this strand where the differences are most apparent—it is in the area of visualisation that, collectively, boys out-perform girls.

Algebra

In some countries, algebra does not appear in the primary school years, coming online when students are moving into secondary schools. In part, this could stem from the Piagetian ideology that primary school students are not able to think abstractly and thus, since algebra is highly abstract, it is seen as being beyond the scope of younger students. However, algebra is about patterning, and being able to identify and describe patterns is integral to the mathematics curriculum and lays the foundation for algebraic thinking (English and Warren, 1998).

Working mathematically

Contemporary thinking sees mathematics as more than a collection of knowledge and skills. It is viewed as a way of thinking and working, a new dimension that is being included in most modern curriculum documents. Debate on the inclusion of mathematics as a separate strand is something that teachers need to address. The argument for mathematics to be integrated is founded on the belief that it is very difficult to undertake much mathematics without working mathematically, thus having it as a separate strand would result in considerable doubling up of content. For example, an operation such as multiplication is often raised in a problem-solving context; while the two items are covered simultaneously, the teacher may focus on only one outcome. Not seeing the operation holistically results in considerably more work.

● Diversity in curriculum organisation

While there is considerable diversity in the organisation of mathematics curricula in various states and countries, there is enough synergy among them to talk in general terms. Typically, curriculum documents are broken down into a number of organising concepts. The five areas discussed above are not meant to be comprehensive; rather, they provide some guidelines to what could typically be expected in a strand. Rather than perceive such an outline as a static organiser, it should be seen as a set of organisers that interrelate with each other. For example, multiplication links strongly with area and volume; negative integers link with temperature; money is a measure but also links with decimal understanding (the Number strand). Because documents are working and evolving over time, different manifestations can reasonably be expected, since writers and developers work under very different constraints. As curriculum documents are developed, demands are placed on writers to conform to other expectations—such as teachers who will be using the documents, policy-makers, systems administrators, parents, employers and other lobby groups. The curriculum should be seen as a representation of particular values and interests within the community from which it emerges.

At the end of the first decade of the twenty-first century, Australia developed a new organising framework for the mathematics curriculum, reducing the five strands to three: Number and Algebra; Measurement and Geometry; and Statistics and Probability. The rationale for this organisation was that it would enable greater connectivity between topics within strands. The Australian mathematics curriculum has integrated proficiency strands that, to an extent, envelop but extend a strand of working mathematically. There are four proficiency strands: Understanding; Fluency; Problem-solving; and Reasoning. The *Understanding* strand emphasises the conceptual aspects of the content and *Fluency* emphasises associated skills. *Problem-solving* is about strategies, communicating mathematically and interpretation of problems; and *Reasoning* incorporates processes of justifying, explaining, analysing, and so on. Curriculum organisation is thus changing and shifting, and reflects the changing society in which we live and the social construction of mathematics curriculum documents.

In recent years, the impact of technology on the mathematics curriculum has also been an area of debate. At the simplest level, the question of calculators has been perennial. Advocates of calculators have shown that their effects have been significant, both as a computational aid and as a learning tool (Gowland, 1998; Ruthven and Chaplin, 1997; Taylor, 1992). Opponents of calculators have argued that they reduce students' capacity to complete even simple tasks. The role of graphic calculators in the secondary school has also been debated on much the same grounds. The use of learning technologies such as spreadsheets for graphing has resulted in similar debates. Overall, the research suggests that calculators and other forms of technology can be powerful tools for supporting, enhancing and extending students' mathematical understandings when used appropriately in the classroom.

Integrating curriculum

The mathematics curriculum gains its benefit and purposefulness as a discipline that informs other disciplines. For example, within the social studies curriculum, where students study environments and demographics, there is a constant referral to graphs and charts. Without the contributing knowledge gained in mathematics, students would be restricted in their capacity to interpret this information. One of the major concerns of mathematics education is the fragmentation of knowledge, which leads to teachers and students not being provided with learning experiences that allow them to make connections between concepts. For students, this can result in their perceiving mathematics as a large discipline with a lot of disconnected facts. For teachers, it can result in the planning of many activities so that considerably more time is spent on achieving the same outcomes. Thus it is important to see the connections between the various strands of the mathematics curriculum, the connections to other curriculum areas, and the connections to the world beyond school.

● Relationships with other areas of mathematics

The various strands of the mathematics curriculum should not be seen in isolation from the other strands or sub-strands; their links to other strands or sub-strands should be considered. For example, it is difficult to teach measurement concepts and skills unless students understand other number concepts such as 1000, understand multiplication processes, and have some sense of number. Similarly, the relationships between sub-strands of volume and capacity can be explored (many students are not aware of the relationship between mL and cm^3, for example). One of the disadvantages of breaking knowledge into compartments, or fragments, is that such connections are at risk of not being noted.

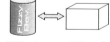

375 mL   375 cm^3

The relationship between liquid and solid measures

Integrating mathematics with other curriculum areas

Many teachers practise using an integrated curriculum model, and subject integration is a signature practice of the middle years of schooling (MYS) pedagogy. This means that they link a number of curriculum areas together. There are three reasons for working in this way:

1. *It works in a pragmatic sense.* With the increasing amount of content to be covered in primary schools, and the increasing demands on teachers' time, teachers have fewer resources available to cover essential learnings. Integration across curriculum areas allows them to ensure that they make the best use of their time. For example, if a teacher is covering report-writing in English, it is possible to link this to science and expect students to write a report for an experiment that has been conducted. Such an experiment may have involved considerable mathematics in measuring results, so links can be made to the measurement strand. Such cross-curriculum links allow teachers to assess and access students' understanding in three areas rather than one.

2. *Mathematics cannot exist on its own.* One of the criticisms of mathematics teaching can be that it is divorced from other areas of learning. Making links to other curriculum areas allows students to see how the skills they are learning in mathematics are relevant to other areas of study. For example, students may be monitoring the water quality in a local creek. This is a part of the environmental science unit. The data that have been collected each day or week can be recorded as tables or graphs and then interpreted to determine the quality of the water. Using their mathematics skills in collecting and recording data, students are able to see their graphing skills as having a real purpose. In another example, students plotting points on a globe require a working knowledge of spherical geometry; thus the application of mathematics is central to being able to identify locations through coordinates. This is a critical aspect of good teaching of mathematics—students are able to see why they need to learn mathematical skills and knowledge.

3. *Mathematics connects to the world beyond school.* Studies on the transfer of mathematics from school to the workplace have

shown very little is in fact transferred between the two contexts. Workers often fail to use the knowledge that they learnt in school mathematics to help them in their work and home environments, and this can restrict the effectiveness and quality of their lives. For example, calculating bank balances or working out the best buys in the supermarket require the knowledge and skills learnt in school; these are experiences in which students can make the important links to the world beyond the mathematics classroom. Being able to use and apply mathematics in daily activities empowers people by enabling them to make informed decisions.

Numeracy

The terms 'numeracy' and 'mathematics' are often used interchangeably. However, numeracy and mathematics are actually very different concepts. Members of the wider community often bemoan the poor 'numeracy' levels of contemporary youth, meaning their inability to calculate. This view of numeracy sees it as being about arithmetic. Some curriculum developers view numeracy as being related to the study of number only. This was very evident in the curriculum reforms in the United Kingdom, where numeracy now focuses on the number strand of the curriculum. A more inclusive definition of numeracy sees it as the knowledge and skills needed for informed participation and decision-making in the world beyond schools (Australian Association of Mathematics Teachers, 1997). This broad definition encompasses the notion of numeracy as being a life skill that is important for living in the contemporary world.

In part, the current use of the term 'numeracy' is politically motivated. Recognising that many young people leave school mathematics with a less than positive disposition towards mathematics, and poor mathematical skills, questions are posed about the value of core school mathematics. Numeracy is therefore a more holistic and applied notion, encompassing expectations that students would use and apply knowledge to solve practical and everyday problems. In contrast, school mathematics would include aspects of numeracy, but also teach abstract knowledge with limited application for most people in their everyday lives.

■ REVIEW QUESTIONS

1.1 What are your experiences of school mathematics? How would you categorise your school mathematics experiences: instrumental or constructionist? Give examples of experiences to support your position.

1.2 How is the mathematics curriculum organised? Write a summary of the key emphases for each strand and/or sub-strand within the mathematics curriculum.

1.3 What is the place of long division in your school curriculum documents? What do the current syllabus documents say about the teaching of long division? What is your view on this?

1.4 What is the place of technology in the school mathematics curriculum? How has technology impacted on the way you learn mathematics? What do new technologies mean for school mathematics?

1.5 A parent comes to your classroom and says that his child needs to be given more pencil-and-paper work, 'where they do lots of sums'. You want to convince him that making the links to the world beyond school is more important than doing a lot of sums on paper. What points could you make to convince him?

Further reading

Historical curriculum documents are a good way to learn more about curriculum organisation, curriculum content and emphases. You should also look at the following:

- *Curriculum documents of your locality.* These may be local, state or national documents. Frequently, authorities will publish curriculum or syllabus documents to which teachers are expected to adhere as they plan their teaching. Beginning teachers need to be familiar with the materials relevant to their contexts.
- *Other support materials.* These can be from commercial publishers and/or can be departmental resource materials developed by local authorities to support teachers in schools.
- *Materials from other regions* (nationally and internationally) so that a broader sense of curriculum can be developed.

- Australian Education Council (1991). *A national statement on mathematics for Australian Schools*. Melbourne: Curriculum Corporation.

References

Anghileri, J. (2001). Intuitive approaches, mental strategies and standard algorithms. In J. Anghileri (ed.), *Principles and practices in arithmetic teaching: Innovative approaches in the primary classroom* (pp. 79–94). Buckingham: Open University Press.

Askew, M., Brown, M., Rhodes,V., Johnson, D. and Wiliam, D. (1997). *Effective teachers of numeracy: Report of a study carried out for the Teacher Training Authority 1995–96*. London: School of Education, King's College.

Australian Association of Mathematics Teachers (AAMT) (1997). *Numeracy = everyone's business; Report of the Numeracy Education Strategy Development Conference.* Adelaide: AAMT.

Beishuizen, M. (1999). The empty number line as a new model. In I. Thompson (ed.), *Issues in teaching numeracy in primary schools* (pp. 157–68). Buckingham: Open University Press.

Bishop, R. and Berryman, M. (2006). *Culture speaks: Cultural relationships and classroom learning.* Wellington, New Zealand: Huia.

Brown, M., Askew, M., Baker, D., Denvir, H., and Millett, A. (1998). Is the national numeracy strategy research-based? *British Journal of Educational Studies,* 46(4), 362–85.

Buys, K. (2001). Progressive mathematisation: Sketch of a learning strand. In J. Anghileri (ed.), *Principles and practices in arithmetic teaching: Innovative approaches for the primary school* (pp. 107–18). Buckingham: Open University Press.

Clements, M.A. (1989). *Mathematics for the minority: Some historical perspectives of school mathematics in Victoria.* Geelong: Deakin University Press.

Education Queensland (2001). *The Queensland longitudinal reform study.* Brisbane: GoPrint.

English, L.D. and Warren, E.A. (1998). Introducing the variable through pattern exploration. *Mathematics Teacher,* 91, 166–70.

Ernest, P. (1991). *The philosophy of mathematics education: Studies in mathematics education.* London: Falmer Press.

Gowland, D. (1998). Calculators: Help or hindrance? *Mathematics in School,* 27(1), 26–8.

Hill, P. W. and Rowe, K.J. (1998). Multilevel modelling in school effectiveness research. *School Effectiveness and School Improvement,* 7, 1–34.

Lowrie, T. (2003). Posing problems in ICT-based contexts. In L. Bragg, C. Campbell, G. Herbert and J. Mousley (eds), *Mathematics education research: Innovation, networking and opportunity. Proceedings of the 26th Annual Conference of the*

Mathematics Education Research Group of Australasia (Vol. 1, pp. 499–506). Geelong: MERGA.

Rosenthal, R. and Jacobsen, L. (1969). *Pygmalion in the classroom*. New York: Holt, Rinehart & Winston.

Ruthven, K. and Chaplin, D. (1997). The calculator as a cognitive tool: Upper primary pupils tackling a realistic number problem. *International Journal of Computers for Mathematical Learning*, 2(93), 93–124.

Taylor, L. (1992). Teaching mathematics with technology. How to win people and influence friends: Calculators in the primary grades. *Arithmetic Teacher*, 39(5), 42–4.

Treffers, A. (1991). Realistic mathematics education in the Netherlands 1980–1990. In L. Streefland (ed.), *Realistic mathematics education in primary school* (pp. 11–20). Utrecht: Freudenthal Institute.

Treffers, A. and Beishuizen, M. (1999). Realistic mathematics education in the Netherlands. In I. Thompson (ed.), *Issues in teaching numeracy in primary schools* (pp. 27–38). Buckingham: Open University Press.

van den Heuvel-Panhuizen, M. (2001). Realistic mathematics education in the Netherlands. In J. Anghileri (ed.), *Principles and practices in arithmetic teaching* (pp. 49–64). Buckingham: Open University Press.

CHAPTER 2

Theories of learning mathematics

I think kids learn maths best when they sit quietly and do lots of practice examples. That way they also know if they are right or not.

I think it is about how I organise the learning for them. Have I pitched just a bit above where they are thinking? Have I posed good questions?

I think it is more about how they construct their own meanings. They all have different experiences which impact on how they make sense of the things we do.

Why study theories of learning mathematics?

To be able to plan how to teach mathematics effectively, there needs to be some understanding of how students learn mathematics. A major UK review of effective teachers of numeracy (Askew et al., 1997) has shown that theory is one of the key factors in developing high-quality practices and outcomes in numeracy learning. The authors argue that 'teachers' beliefs and understandings of the mathematics and pedagogical purposes behind particular classroom practices seems to be more important than the forms of practices themselves' (p. 3). It is not so much the ways in which teachers use particular practices in the classroom (e.g. whole group work, mental maths, direct teaching or other methods) that are critical, but rather the beliefs they hold towards their teaching approaches. The role of theory in underpinning practice is an essential element of quality teaching in mathematics.

By having an idea of how students learn, teachers are better able to plan and anticipate in particular ways, and to create learning environments to facilitate better learning. Three significant classes of theory have had a strong influence on our understandings of how students come to learn and understand mathematics:

1. cognitive theories that focus on students' thinking
2. sociocultural theories that seek to understand cognition within a social context
3. social (or socially critical) theories.

This chapter provides an overview of key ideas associated with these theories. Theories associated with learning in 'modern society' are also discussed.

Cognitive theories

● The influence of Piaget

The work of Jean Piaget (1972) has had a significant impact on mathematics education. His work on two aspects of how ideas were formed spanned a considerable part of the twentieth century. His early and late periods were dominated by the active construction of meaning, where he proposed that, through the twin processes of accommodation and assimilation, schema were constructed. These ideas are foundational to the significant impact of constructivism in mathematics education. Piaget's middle period was dominated by stage theories, through which he tried to develop an idea of how young students went through particular stages in their thinking patterns. Piaget's writings on children's cognitive and affective development were extremely influential in education, particularly from the 1950s onwards. His stage theory of cognitive development, for example, strongly influenced early childhood and primary education in the 1960s and 1970s. In more recent years, the influence of this theory has declined because of the view that it tends to highlight what children *cannot* do, rather than what they *can* do. As well, stage theory, in the forms in which it has been translated (from French), interpreted and applied, is criticised on the basis that it highlights notions of readiness. In doing so, it can hold back particular aspects of instruction at the expense of teaching particular content.

One important focus of Piaget's work, highly relevant to our contrast, is the development of logico-mathematical knowledge. In particular, he made a major contribution to the understanding of the development of number concepts in young children as well as the development of concepts relating to logic, time, space and geometry, and movement and speed. Since the early 1980s, Piaget's theories have constituted one of the main bases for the development of constructivist theories in education.

● Constructivism

'Constructivism' is a term that has been used in education and educational psychology with increasing frequency since the late 1970s. Today, any serious discussion of learning theory related to mathematics, science or literacy, for example, would include a detailed discussion of constructivism. As outlined by Cobb (2000): 'A range of psychological theories about learning and understanding falls under the heading of constructivism. The common element that ties together this family of theories is the assumption that people actively build or construct their knowledge of the world and of each other.' (p. 277) Cobb describes how constructivists reject notions of stimulus–response theory (behaviourist learning theory) and how remembering is more than direct retrieval, as thought processes are in operation. The individual mind of the learner is central to constructivism.

The impact of Piaget's work in contemporary mathematics is obvious in the ways in which constructivism, in its numerous versions, has been taken up by teachers and curriculum writers. There are a number of different forms of constructivism, but underpinning all versions are three premises:

1. Rather than being passively received, knowledge is actively constructed by students.
2. Mathematical knowledge is created by students as they reflect on their physical and mental actions. By observing relationships, identifying patterns and making abstractions and generalisations, students come to integrate new knowledge into their existing mathematical schemas.
3. Learning mathematics is a social process where, through dialogue and interaction, students come to construct more refined mathematical

knowledge. Through engaging in the physical and social aspects of mathematics, students come to construct more robust understandings of mathematical concepts and processes through processes of negotiation, explanation and justification.

Constructivism recognises that mathematics must make sense to students if they are to retain and learn mathematics. For students to develop appropriate knowledge, they must be provided with rich learning experiences so that their constructed meanings and understandings are in keeping with the discipline of mathematics.

● The importance of dialogue and argumentation

Within the constructivist paradigm, the role of language and dialogue is central to fostering learning environments. Providing appropriately organised experiences where students can talk with their peers allows them to explore ideas in language and concepts similar to their own. This enables higher-achieving students to practise their control of language and lower-achieving students to hear ideas being modelled in a language that is more likely to be in a genre that they can access. For example, when talking about the properties of three-dimensional shapes, the formal language of edges, faces and vertices may be introduced. The high-achieving students may find this language useful as they have been confused by the use of the term 'side'—does it refer to the face or an edge? The appropriate language aids their learning. Other students may be grappling with notions of three-dimensional shapes. Having students explain their ideas to their peers often supports both sets of learners.

● Constructivism in the mathematics classroom

Within a constructivist classroom, the teacher acknowledges that students will have constructed a range of understandings from any given interaction on the basis that they have entered the context from a range of different perspectives and experiences. A constructivist perspective recognises that it is not possible to assume the teaching of a concept relates to the development of the ideas proposed by the teacher, and that there will be a multiplicity of understandings constructed by the range of students in the classroom. A constructivist teacher realises that having taught something does not mean that students have learnt exactly what was envisaged by the teacher. It is important for the teacher to use a range of tools and techniques to assess what the students have constructed.

By identifying what the students have constructed, the teacher is then able to identify constructions akin to the objective of the lesson, as well as to identify misconceptions. It is the misconceptions that allow the teacher to access what the students have come to construct, and thus develop teaching strategies that will move the students into more appropriate constructions.

Sociocultural theories: The influence of Vygotsky

Lev Vygotsky is regarded as the founder of sociocultural theory, or what can be described as the sociohistorical approach in psychology (e.g. Cole, 1996; Moll, 1990). Vygotsky's work, which is embodied in the literature on sociocultural theories of learning mathematics, has gained increasing importance in theorising how students learn mathematics. Vygotsky saw that students internalised complex ideas (Daniels, 1990), but he extended the general constructivist approach by arguing that the internalisation of knowledge could be achieved more effectively when students were guided by good, analytic questions posed by the teacher.

The expert teacher is central to Vygotskian theory. The teacher's role is to identify the student's current mode of representation and then, through the use of good discourse, questioning or learning situations, provoke students to move forward in their thinking. The recognition of a student's representation or thinking was seen as their zone of proximal development and the teacher's actions to support learning were described as scaffolding. When working in the 'zone of proximal development', particular attention is paid to the language being used since the language of students influences how they will interpret and build understandings (Bell and Woo, 1998). Within a Vygotskian approach, it is seen to be important that teachers use and build considerable language and communication opportunities within the classroom environment in order to build mathematical understandings.

● Scaffolding

Good teaching involves teachers knowing their students' current thinking about mathematical concepts and then understanding how to move the students towards more complex, complete and robust constructions through the use of organised learning activities and environments. Good questions are important in facilitating learning. Typically, good questions are those that foster deeper levels of learning as opposed to simple recall.

Socially critical theories

Sociological theories, and particularly critical sociology, have gained increasing importance in mathematics. These theories shift the focus of learning mathematics away from the individual to a more macro level of analysis. In part, this interest has stemmed from the consistently poor performance of students who come from particular backgrounds. It is now recognised internationally that particular students are more at risk of not performing well. Aside from students with learning disabilities, these are indigenous students of almost all nations, students from working-class (or low socioeconomic status) families, students who live in remote or rural areas and students whose first language is not English (in English-speaking countries). When gender is considered in concert with these variables, differences are exacerbated (Walkerdine, 1988, 1989). It has long been recognised that girls in particular have been disadvantaged in the study of mathematics as a result of gendered practices in teaching and assessment (Fennema and Meyer, 1989; Leder, 1992). However, it is now recognised that this applies not to girls *per se*, but rather to girls from particular social and cultural backgrounds. When considering the ways in which the practices of mathematics work to exclude girls, it becomes important to recognise that some girls (generally middle-class girls) are more likely to be successful than their peers from working-class backgrounds (both girls and boys); thus gender is not the sole variable, but must be considered in concert with other variables. Rather than assume that success is due to some innate 'mathematical ability', socially critical theories focus their attention on the practices of school mathematics.

Socially critical theories explore the practices of mathematics education to see how they are implicated in the reproduction of inequities, and in so doing challenge such practices to change. Assessment, mathematical language (Zevenbergen, 2000, 2001) and classroom talk, textbooks (Dowling, 1998) and ability grouping, (Boaler, 1997) are some of the areas that have been critically examined in terms of the ways in which social, cultural, linguistic and gender differences are reproduced through mathematics education. These studies have illustrated the very subtle ways in which school mathematics contributes to, and legitimates, the failure of particular groups of students. This book devotes a full chapter to the study of equity from this perspective (see Chapter 4).

Mathematics is one of the most important subjects in the school curriculum, and it serves a particular role (among others) as a social filter. The work of Lamb (1997) shows that success in school mathematics is the best predictor of success in life. People who come to believe that they are not good at mathematics tend to accept their position in life.Thus it is important for all students to succeed in school mathematics—regardless of background, gender or language. By knowing how practices are implicated in the construction of differences, teachers can change their practice in order to produce more equitable classrooms and outcomes.

Modern society

Current educational theory has moved to a stronger recognition of how society impacts on learning. Postmodern theorising has drawn considerable attention to how current times are very different from old times (the Industrial Age), largely due to the use of technology. As early as 30 years ago, educationalists were commenting on how television was influencing attention rates and suggesting that young students had much shorter attention spans than in the past. Similarly, the advent of television meant that reading books was being replaced by viewing television, and there was seen to be a consequent decline in reading skills and motivations. More recently, the use of computers and other computer technologies has been seen to have considerable influence over students' thinking and behaviours, so much so that terms such as 'cyberkids' and 'technoliteracy' have become part of the educational discourse (Luke, 2000).

The world for which schools are preparing students is fundamentally different from that of even a few years ago. The term 'modern society' has been coined to represent the very different social, economic, political and educational times of the society in which young people now live (Gee, 2002). Many of the new terms in educational theory are prefaced with the word 'new' to represent this thinking. Students who have grown up in a technological age are less intimidated by technology, and hence its insertion into mathematics is an important change.

● Technology and 'modern society'

In terms of learning, 'modern society' embraces the use of technology so that the tedium often associated with working calculations ('doing sums') can be replaced with technologies to support the development of mathematical thinking. This change is fundamental in preparing students for life in modern society. Technology in this sense includes calculators as well as computers, so that ways of thinking—such as the development of algebraic thinking—can be very well supported by the use of technology (Asp and McCrae, 2000).

Technology and old curriculums have been challenged by the work of Stacey and Groves (1996). In an old curriculum, students in the first year or two of schooling would only work with numbers to 20. Through the use of calculators, young students' number sense can now be developed to numbers with four or more digits (Groves, 1995). Thus, rather than technology replacing skills, it can be used to enhance mathematical thinking.

Within this framework, consideration must also be given to what is seen as mathematics. Often this is framed within basic skills, with the wider society bemoaning young graduates who are unable to calculate. Within modern society, the emphasis becomes somewhat different due to the saturation of technology. Thirty years ago, shops could only enter the value of the goods in the cash register—calculations were done on paper, and change given using a counting-on method. Today's society is much richer in technology—some stores have registers that scan items, so no data need be manually entered, while others require the assistant only to press the item of purchase (e.g. McDonald's). Most registers also calculate the amount of refund (change) required. Thus old basics have been superseded by new basics—sales assistants need to be able to estimate, to problem-solve, to validate, in order to evaluate whether items have been scanned or not and thus verify the validity of the total amount or the change given (in case a wrong amount is entered for amount tendered). In other words, they need some of the old basics, but technology has brought with it a range of new skills. These skills are different from, but not exclusive of, the attributes that would be sought in mathematics.

Within a modern society framework, the teaching of mathematics involves a much greater emphasis on the use and integration of learning technologies (such as computers, calculators and graphic calculators). Since most Western students have been exposed to such technologies, they are more likely to think and respond to these forms of teaching.

Furthermore, there is a renewed emphasis on the mathematics curriculum containing new forms of mathematics that reflect the needs of the world beyond. Modern society reflects the supersaturation of information, so students need to be more 'literate' in terms of analysing, interpreting and being able to critique the texts to which they are exposed. This often means being able to use and apply their mathematical thinking and analysis to texts containing data—such as graphs, measures of central tendency, and so on. They not only need the skills required to construct or calculate such measures, but within modern society they also need to be able to interpret such information more carefully. This demands a refocusing of the mathematics curriculum into such areas.

Theory into practice

The value of a good theory is its capacity to enable teachers to develop good practice that supports and enhances student learning. Teachers need to have a strong theoretical basis to their work. By understanding how students learn, teachers are able to organise the learning in ways that will enhance the capacity for learning. Rather than advocate one theory as being superior to another, it is more appropriate to consider what is being learnt.

■ REVIEW QUESTIONS

2.1 What are the main characteristics of the constructivist theories underpinning mathematics teaching and learning?

2.2 Discuss how you would organise learning when using scaffolding. Provide an example.

2.3 Describe how you would teach coordinates and outline the theories of learning that you would draw upon.

2.4 In what ways does a socially critical approach to learning mathematics differ from a cognitive or sociocultural approach? Outline the main tenets of such approaches.

2.5 List the challenges faced by teachers in the teaching of mathematics in modern society.

Further reading

Mousley, J. (2001). Theories of learning: What's next? From construction to activity theory. *Vinculum*, 39(3), 8–13.

Reys, B. and Long, V.M. (1995). Teachers as architects of mathematical tasks. *Teaching Children Mathematics*, 1(5), 296–9.

Rowan, T.E. and Robles, J. (1998). Using questions to help children build mathematical power. *Teaching Children Mathematics,* 4(9), 504–9.

Stephens M., Montgomery, P. and Waters, M. (1997). Creating a climate for effective project work in mathematics. *Australian Primary Mathematics Classroom,* 2(2), 6–10.

References

Askew, M., Brown, M., Rhodes,V., Johnson, D. and Wiliam, D. (1997). *Effective teachers of numeracy: Report of a study carried out for the Teacher Training Authority 1995–96.* London: School of Education, King's College.

Asp, G. and McCrae, B. (2000). Technology-assisted mathematics education. In K. Owens and J. Mousley (eds), *Research in mathematics education in Australasia 1996–1999* (pp. 123–60). Sydney: Mathematics Education Research Group of Australasia.

Bell, G. and Woo, J.H. (1998). Probing the links between language and *mathematical conceptualisation. Mathematics Education Research Journal, 10(1),* 51–74.

Boaler, J. (1997). *Experiencing school mathematics: Teaching styles, sex and setting.* Buckingham: Open University Press.

Cobb, P. (2000). Constructivism. In A.E. Kazdin (ed.), *Encyclopedia of psychology* (Vol. 2, pp. 277–9). Washington, DC and New York: American Psychological Association and Oxford University Press.

Cole, M. (1996). *Cultural psychology: A once and future discipline*. Cambridge, MA: Harvard University Press.

Daniels, H. (1990). Number competence and communication difficulty: A Vygotskian analysis. *Educational Studies*, 16(1), 49–59.

Dowling, P. (1998). *The sociology of mathematics education: Mathematical myths/ pedagogical texts* (Vol. 7). London: Falmer Press.

Fennema, E. and Meyer, M.R. (1989). Gender, equity and mathematics. In W. Secada (ed.), *Equity in education*. New York: Falmer Press.

Gee, J. P. (2002). New Times and new literacies: Themes for a changing world. In B. Cope and M. Kalantzis (eds), *Learning for the future: Proceedings of the learning conference* (pp. 3–20). Spetses, Greece: Common Ground.

Groves, S. (1995). The impact of calculator use on young children's development of number concepts. In R.P. Hunting, G.E. Fitzsimons, P.C. Clarkson and A.J. Bishop (eds), *Regional collaboration in mathematics education* (pp. 301–10). Melbourne: Monash University.

Lamb, S. (1997). *Longitudinal study of youth labour markets*. Melbourne: Australian Council of Educational Research.

Leder, G. (1992). Mathematics and gender: Changing perspectives. In G. Grouws (ed.), *Handbook of research on mathematics teaching and learning: A project of the National Council of Mathematics Teachers*. New York: Maxwell Macmillan.

Luke, C. (2000). Cyber-schooling and technological change: Multiliteracies in New Times. In B. Cope and M. Kalantzis (eds), *Multiliteracies: Literacy learning and the design of social futures* (pp. 244–66). Melbourne: Macmillan.

Moll, L.C. (1990). *Vygotsky and education: Instructional implications and applications of sociohistorical psychology*. Cambridge: Cambridge University Press.

Piaget, J. (1972). *To learn is to invent*. New York: Grossman.

Stacey, K. and Groves, S. (1996). Redefining early number concepts through calculator use. In J. Mulligan and M. Mitchelmore (eds), *Children's number learning* (pp. 205–25). Adelaide: Australian Association of Mathematics Teachers.

Walkerdine, V. (1988). *The mastery of reason: Cognitive development and the production of rationality*. London: Routledge.

——(1989). *Counting girls out*. London: Virago.

Zevenbergen, R. (2000). 'Cracking the code' of mathematics: School success as a function of linguistic, social and cultural background. In J. Boaler (ed.), *Multiple perspectives on mathematics teaching and learning* (pp. 201–23). New York: JAI/Ablex.

——(2001). Mathematics, social class and linguistic capital: An analysis of a mathematics classroom. In B. Atweh and H. Forgasz (eds), *Socio-cultural aspects of mathematics education: An international perspective* (pp. 201–15). Mahwah, NJ: Lawrence Erlbaum.

CHAPTER 3

Language and mathematics

For a long time, mathematics was seen as a subject area that remained autonomous from language. This view has been challenged considerably over the years, to the point where it is now recognised that language plays a central role in mathematics. Students need to learn the language of mathematics in order to participate in and learn the discipline. Mathematics has a very particular language, and being able to read, interpret and respond in that language are all central to being an effective learner.

The language of mathematics

Typically, coming to learn and understand mathematics involves a pairing of written words, oral sounds or abstract symbols with particular meanings. For example, the word 'ruler' is used to refer to a measuring instrument usually used to measure lengths. Similarly, the '+' sign is used to refer to the addition concept/process where, in most cases, two sets are combined to form a superset. In learning mathematics, students need to learn the particular relationships between signs and meanings.

Many new words are introduced, some of which will be familiar, some of which are unique to mathematics. To be competent and effective in mathematics, students need to learn this language. While some words may create difficulties due to their linguistic complexity (e.g. tessellation) or the concept to which they refer (e.g. parallel), others create problems because they are familiar from everyday discourse (e.g. irrational, ruler, odd) but take on a specific and/or different meaning in mathematics.

When teaching mathematical ideas, teachers need to be sure that students understand the very particular language of mathematics. Often what are seen as misconceptions emerge due to the use of lay terms and meanings. When working with fractions, the numerator is often interpreted as the 'top number' and the denominator as the 'bottom number'. This 'misconception', while working well with fractions, cannot be transferred to other contexts such as subtraction, where we have heard teachers say 'the numerator is taken away from the denominator'. This is incorrect and shows the misappropriation of language.

● The development of a specialised mathematical language

Part of the challenge in understanding mathematics is learning the specific language of mathematics. This may mean learning complex mathematical terms; however, the more difficult areas for students are when terms familiar to them in everyday language are used in more specific ways in mathematics. For example, consider the student who does not like 'odd numbers because they are strange'. Here the student transfers the everyday meaning of 'odd' to the mathematics context, resulting in a novel interpretation of a mathematical idea. Other words that can lead to confusion include ruler, mass, root and vulgar.

Even within mathematics, the same word can be used in different ways. Part of learning mathematics is the progressive refinement of language. For example, when working with shapes, the common words 'sides' and 'corners' are used in the early years. But this becomes problematic when three-dimensional shapes are introduced. 'Sides' on the two-dimensional shape is used early to draw attention to the lines that make up the shape, but when talking about three-dimensional shapes the term 'side' can become confusing—does it refer to the line or the shaded area? Similarly, the term 'corner' becomes problematic.

Where is the side?

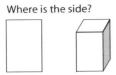

Where is the corner?

As a result, the very specific language of shape needs to be introduced. Learning the language of mathematics can be very empowering for students as they progress through school, and teachers need to pay particular attention to teaching the language of mathematics.

In coming to communicate mathematically, students need to understand the language of mathematics. In being aware of potentially problematic areas, teachers will better be able to organise effective learning environments for all students. This is particularly important when considering students whose first language is not that of instruction. Teachers need to remember that this includes English-speaking students whose language is different from the middle-class English normally used in the classroom.

● Ambiguity in terminology

Mathematical terms that are ambiguous in meaning (i.e. have multiple meanings) can create difficulties for students. Imagine the confusion of the young students who hear about the exciting lesson about to begin on rulers. Thinking that they are going to hear about kings and queens, princes and princesses, emperors and empresses, in the first part of the lesson the students will be awash with confusion as the teacher brings out a 30 cm piece of wood and talks about it. The confusion, and loss of learning, has occurred due to a misinterpretation of the term 'ruler'.

Three types of words fall into this category of ambiguity: homonyms, homophones and polysemes.

Homonyms: Shifting between the everyday and school mathematics

Homonyms are words that look and sound the same but have different meanings—such as mass, vulgar, irrational and odd. Students can become confused when they hear one meaning but the teacher is intending the term to be interpreted in another way. An example of this can occur when students are asked to 'calculate the *volume* of a box'. Some students will interpret this as a three-dimensional task, while others will become confused as they wonder how they will calculate the 'noise' of the box. Teachers need to be aware of the homonyms used in a lesson to avoid confusion.

This area can be particularly confusing when spatial terminology is tied in with number study. In the examples here, students were asked to write a 'high' number and a 'big' number. High and big were meant to infer three-digit numbers but the students interpreted them as spatial rather than in terms of magnitude (see diagram).

6	
	3
high number	big number

There is considerable potential for other areas of mathematics language to create confusion—consider the question: 'What is the length of time between 9.05 a.m. and 9.55 a.m.?' How do students make sense of this type of question when they have a construct of length as a unit of measure that can be measured with a ruler?

Homonyms common in mathematics

angle	face	odd	right
average	figure	parallel	root
base	improper	point	sign
below	leaves	power	similar
cardinal	left	product	square
change	make	proper	table
common	mean	rational	times
degree	model	real	unit
difference	natural	record	volume

The list of homonyms given in the table does not include every homonym found in mathematics; it merely highlights some of those most commonly used. When planning lessons, teachers should always consider the language being used in a lesson, with specific attention paid to these types of words. When they appear in the teaching episode, teachers should overtly draw attention to the specific meaning being used in the immediate context.

Homophones

Whereas homonyms are the same word with different meanings, homophones have different spellings but the same sound. In teaching the addition of fractions, it is common to hear phrases such as 'two halves make a whole'. In a lesson where two halves are being placed together to make a full item, and the student is hearing 'two halves make

a hole'—but no hole is being created with the combining of the two objects (unless a doughnut is being used as an example, which would develop further misconceptions!)—the student could be left wondering what the teacher is talking about. This results in considerable talking past the student, who is unable to make sense of what is happening in spite of expending much effort to understand.

Common homophones

for—four	pie—pi
hole—whole	route—root
one—won	to—too—two

When terms such as those in the table appear in a teaching episode, potential confusion can be avoided by writing both words on the board. Since the words are different in appearance (spelling), students can then see the differences.

Polysemy

The third type of confusing word covers those that have two or more related meanings. The term 'product' is an example of a polysemous word. Similarly, students can become confused by the term 'base', which means foundation. Having developed a construct of base as the bottom of a shape (such as a pyramid), number study then brings in a different idea of base, as in base 100. Students will not necessarily make the connection between the two meanings if they do not have a sufficient grasp of language to see base as a foundation.

Clearly, where lexical ambiguity exists, teachers need to develop an acute awareness of the language they will use in a lesson. When planning a teaching episode, language should be an important consideration—both in terms of the new language to be learnt and any language that may become problematic. As discussed in Chapter 5, language is a key consideration in the overall planning of a lesson.

● Prepositions: The little words that mean a lot

Mathematics language is very specific and very concise. The types of reading skills that students develop as they progress through school may mean that many of the 'little' words are skipped over. This can become a difficulty in mathematics since there is little redundancy in mathematical language. Students need to read mathematical texts very carefully, especially with regard to prepositions. Prepositions often have a very specific function in mathematics. For example, 'off ' and 'of ' are small words, but they can affect responses significantly. When working through percentages, '25% *of* a price' as opposed to '25% *off* a price' results in two very different answers. Paying attention to this level of detail is important.

Students develop strategies where they search for key information in other subjects and use this to solve their tasks, but they often pay little attention to the prepositions. In mathematics, this strategy can be flawed. It is particularly problematic for students whose command of English is not strong and where prepositions are not easily learnt. In teaching students to read mathematically, the teacher needs to draw attention to the words being used and to encourage students to read mathematical texts differently (and more carefully) than non-mathematical texts.

Word problems

Within the literature on language and mathematics, a significant body of research has been undertaken with word problems. In attempting to link mathematics to the world beyond schools, many arithmetic tasks have been inserted into worded problems. For example, $3 + 4 =$ becomes 'If I went shopping and bought three apples and four oranges, how many pieces of fruit would I have?' Part of the reasoning behind this is to make mathematics more realistic. However, it can create another layer of complexity for students through the language of the problem.

One of the problems with mathematics teaching and learning is that students come to see school mathematics in particular ways.

They develop a sense of the culture of the classroom in which answers are expected. They also learn strategies for reading word problems. Most of these strategies work well for the standard questions being posed. The work of Freebody (1992) illustrates the importance of the roles of the literate student, which are very evident in the mathematics classroom. The ability to decipher, interpret and respond to word problems effectively is an important skill for students to acquire. For example, when posed a nonsensical problem such as 'There are 60 ice-cream flavours. If the shopkeeper has $20 in his pocket, how much would an ice cream cost?', students may attempt to answer the question by playing with the numbers and not addressing the task. This strategy of taking key words and numbers, then trying to provide an acceptable answer, is common in mathematics classrooms, and highlights the inappropriate literacy strategies employed by students. A greater emphasis on reading for meaning (as opposed to 'doing') needs to be an integral part of the teaching of mathematics. Using the literacy model, a competent reader of a mathematical text is one who is able to 'break the code' to make meaning of the text so as to use the text in an appropriate manner and critically analyse the text (Freebody, 1992). These elements apply to the types of literacy demanded in mathematics classrooms.

● Heller and Greeno classification of word problems

Within the word problems posed in the mathematics classroom, a particular form of questions is often used. (Other word problems are discussed later in the chapter.) Heller and Greeno (1978) classify and order word problems based on the actions undertaken and the level of difficulty, and have proposed a taxonomy of word problems. They identify three main types of problem—change, combine and compare.

1. *Change questions* involve a dynamic process whereby there is an event that alters the value of the quantity—for example, Michael had five apples; Kelly gave him two more apples. How many apples does Michael have now?
2. *Combine questions* relate to static situations where there are two amounts. These are considered either as separate entities or in relation to each other—for example, Steven has four oranges; Michelle has two oranges. How many oranges do they have altogether?
3. *Compare questions* involve the comparison of two amounts and the difference between them—for example, Stephanie has five oranges; Alice has two more oranges than Stephanie. How many oranges does Alice have?

The arithmetic is very simple, but there is complexity due to the semantic structure of the problems. The levels of difficulty depend on the operation to be undertaken (addition or subtraction), the type of word problem (change, combine or compare) and the location of the unknown set.

Problems involving addition are usually solved more frequently than similar problems involving subtraction. Similarly, where the unknown set is the solution, this is solved more frequently than when the set is at the start of the problem. These trends have been confirmed across numerous studies (Heller and Greeno, 1978; Lean et al., 1990; Verschaffel and de Corte, 1997; Zevenbergen et al., 2001). Some studies have found that simply rewording the problems makes them more accessible to students (Davis-Dorey et al., 1991; de Corte et al., 1985).

With these sorts of problems, the demands on students are not highly complex insofar as the arithmetic is concerned; the complexity comes about through the structuring of the questions. The complex organisation of the sentences and the order of the operations to be undertaken are potentially confused. In most cases, the examples of word problems students encounter in classrooms present as clear problems where they do not have to rely on deciphering the linguistic demands of the task. Rather, they rely on cues that they build up from earlier experiences—for example, with the commonly posed problem form, 'I had six apples, Jenny gave me three more, how many did I have altogether?' Students have had numerous experiences with this type of question so they can generally rely on a method where they seek out key words—such as 'more' and 'altogether'—and assume that the task requires addition. However, what has been lost in this process is a textual or literacy component where the students need to read and work out what the question is actually demanding. When posing problems of this kind to students, teachers need to be aware that the arithmetic may not be difficult but the *structuring* of the problem may cause difficulties. Teachers need to recognise the ways in which these types of word problems are structured, and recognise any inherent complexity in the language used.

● Realistic word problems

Another form of word problem occurs in attempts to embed mathematics into realistic (or real-world) contexts. There is some debate as to whether or not such questions are real-world ones. Problems of this format look something like 'A farmer has 6 hectares of prime land. It is divided into four equal sized paddocks. If he can graze six cows per hectare, how many cows can he put in each paddock?' In these problems, the teacher has tried to embed the mathematics into a context that appears to be a realistic one—a task of division into a farming context. In part, this is to show the rationale for doing this type of arithmetic.

In order to complete these tasks, students need to be able to decipher the text for what is being asked and then to respond in an appropriate manner. In the following section, a tool is provided for working through the errors that students make in responding to these tasks. However, sociological theories of school performance indicate potential difficulties for students about which teachers need to be aware when posing this type of question. Consider the following question: 'There are 315 students at the sports carnival. If a bus holds 60 people, how many buses are needed to get all the students back to school?' Within a mathematics classroom, the students are expected to divide 315 by 60, find that there is a remainder and then round up to the next whole number. The answer 6 would be seen as the desired response. However, some students may offer answers of '5 rem 15', which indicates that they did not realise there was a need to round up. Others may offer responses of '5 and put some kids so there are 3 per seat' or '5 and some kids go home with the teachers in their cars'. The last two responses suggest that the students have not read the context of the problem correctly and interpreted it as a real-world problem (as opposed to a school problem) and responded within that discourse.

When using real-world problems such as those described above, teachers need to be aware that there is layering of complexity. Not only do students need to have the literacy skills to decode the question in order to respond, but they also need to be able to critically appraise the context of the question in order to decide that the question is really a mathematics task masked in an everyday context, and that the correct response is one that is embedded in a mathematics discourse as opposed to an everyday discourse. The correct identification of the appropriate discourse in which to reply may require teachers to undertake explicit instruction in the language game that is being lived out in such teaching episodes.

Language error analysis

Many of the tasks set in mathematics classrooms have been embedded in some word-based problem where students need to read the problem and work out a response. When a student makes an error, there are a number of levels at which the error could be made that do not necessarily represent a mathematical error. Consider the problem posed in the previous section: 'There are 315 students at the sports carnival. If a bus holds 60 people, how many buses are needed to get all the students back to school?' The work of Newman (1983) provides insights into this perspective. In her work on error analysis, and its adaptations (Ellerton and Clements, 1992), she shows that there are five levels to be undertaken before a student is able to offer the desired response of six buses. Newman and her advocates propose that students:

1. must be able to read the question (can they read the wording?)
2. comprehend what the task is asking (can they identify the task as asking how many buses are needed to transport students?)
3. need to be able to translate this into the mathematical demands (can they identify that the task involves division and rounding up?)
4. undertake the necessary mathematical operations in order to calculate or whatever process is demanded by the task (can they

divide 315 by 60 and then round this to the next highest whole number?) and

5. represent the answer as a meaningful construct (in this case, six buses).

Frequently, incorrect responses have been assumed to be a result of the student not being able to do the mathematics—since the problem is a mathematical one. In contrast, this model indicates that there are both literacy and mathematical demands within the tasks. These five levels, or steps, indicate different areas where students can make mistakes.

Teachers need to be aware that the production of an incorrect response may be due to an error at any of these steps. In considering the bus example, Newman's error analysis gives educators a method for deciphering where students may be making errors when they undertake these sorts of task. It moves considerably away from a view that the errors are purely in the mathematics, indicating there are other sites for errors—namely in reading and comprehending the task. It therefore indicates both literacy and numeracy demands for a given task.

This is a very useful diagnostic tool for teachers to decipher how and where students may be making errors (aside from those that are due to carelessness). It is helpful in providing a sequence through which teachers can track progress in learning and comprehending. By posing relevant questions, teachers are able to identify the level at which the students are experiencing difficulties. The first three levels can be seen to relate to literacy demands whereas the last two tend to be more mathematical. What this model demonstrates is that placing arithmetical tasks into contextual problems places a greater demand on students—particularly in terms of readability—and that this, in turn, creates another level of difficulty for them.

Mathematical literacies

The term 'literacy' is one that causes considerable debate in terms of its depth and application. The term 'multiliteracies' has gained some recognition in attempting to capture this debate so that the particular and more contemporary forms of literacy needed for the emerging societies of the twenty-first century can be recognised (Unsworth, 2002). For example, the literacies needed in the digital environment, the literacies needed for communication within various social, work, educational or other contexts, and the literacies needed for reading and comprehending modern forms of communication such as media and television (Luke, 2000) can all be encompassed within the more traditional forms of literacy. These new forms of literacy are very evident in the teaching and learning of mathematics where students are expected to be computer literate in order to undertake successful engagement in activities related to various software packages, including spreadsheets (West, 2000) and specialised geometry packages (Cabri, or even graphics calculators). When it comes to reading maps, students have developed new forms of digital map-reading through the electronic games that many play (Lowrie, 2003).

Within the electronic games format, students only have access to the immediate screen but are able to visualise far more screens than are seen and to predict movements in these non-seen spaces. Another form of literacy involves appropriate ways of acting and responding in the classroom. As the mathematics classroom is a particular social and cultural context, students need to be able to identify appropriate ways of reading and interpreting social situations in order to be effective participants in the classroom.

In teaching students to apply mathematics to other contexts and situations, teachers often embed mathematics tasks into real-world problems—such as the bus example cited earlier. Textbooks and teacher resources abound with these examples. Two levels of literacy are needed in order to respond to these tasks. The first is the functional literacy noted earlier, where students need to be able to read and decode the text. The second level is where students need to develop a social literacy in order to decipher the unspoken rules of the game in order to respond to such tasks. Often students misinterpret questions as 'trick' questions, whereas it is more likely that they did not correctly interpret the unspoken rules of the game.

Making pedagogy explicit

In a large study of classroom-based research, Sullivan and colleagues (Sullivan et al., 2002) have focused on teachers making aspects of pedagogy explicit to students. Many of the issues identified in this chapter are not taught explicitly to students. They are expected to be learnt through a process of osmosis. This results in some students cracking the code of classroom literacies and others making little sense of the game being played. Unfortunately, a failure to 'crack the code' can have serious consequences. For ethical teaching, teachers need to be aware of the ways in which language is implicated in teaching mathematics, and make the game explicit to their students. This can be undertaken in relatively simple ways—for example, by writing new words on the board or by explaining that the game being played is mathematics (rather than a pseudo-real context). Using the strategies of literacy and English as a Second Language (CESL) classrooms, teachers can support their students in learning the language of mathematics—and its multiple forms of literacy. These are simple strategies, but they can have profound effects for learners.

■ REVIEW QUESTIONS

3.1 What are some of the key language issues that can create difficulties for students? Provide examples for each one.

3.2 In terms of test items, what are some key areas that can create difficulties for students?

3.3 How might you go about teaching language explicitly in the mathematics classroom?

3.4 Reflect on your teaching experiences and identify an example where students had difficulties in responding to your questions due to recognition and realisation rules.

3.5 Using Newman error analysis, work with a student to identify where they may be making mistakes when responding to word problems.

3.6 There are three different types of word problem. List each one and provide an example, outlining why it belongs to the category.

Further reading

Ellerton, N.F. and Clements, M.A. (1991). *The mathematics of language: A review of language factors in school mathematics*. Geelong: Deakin University Press.

Orton, A. (1996) *Learning mathematics: Issues, theory and classroom practice* (2nd ed.). London: Cassell.

Pimm, D. (1991). Communicating mathematically. In K. Durkin and B. Shire (eds), *Language in mathematical education: Research and practice* (pp. 17–24). Philadelphia, PA: Open University Press.

References

Davis-Dorey, J., Ross, S.M. and Morrison, G.R. (1991). The role of rewording and context personalization in the solving of mathematical word problems. *Journal of Educational Psychology*, 83(1), 61–8.

de Corte, E., Verschaffel, L. and De Win, L. (1985). The influence of rewording verbal problems on children's problem representations and solutions. *Journal of Educational Psychology*, 77, 460–70.

Ellerton, N.F. and Clements, M.A. (1992). Implications of Newman research for the issue of 'What is basic in school mathematics?' In K. Owens (ed.), *Space: The first and final frontier* (pp. 276–84). Sydney: Mathematics Education Research Group of Australasia.

Freebody, P. (1992). A socio-cultural approach: Resourcing the four roles of the literacy learner. In A.J. Watson and A.M. Badenthorp (eds), *Preventing reading failure*. Sydney: Ashton Scholastic.

Heller, J.I. and Greeno, J.G. (1978). Semantic processing of arithmetic word problems. *Paper presented at the annual meeting of the Midwestern Psychological Association*.

Lean, G.A., Clements, M.A. and Del Campo, G. (1990). Linguistic and pedagogical factors affecting children's understanding of arithmetic word problems: A comparative study. *Educational Studies in Mathematics*, 21, 165–91.

Lowrie, T. (2003). Posing problems in ICT-based contexts. In L. Bragg, C. Campbell, G. Herbert and J. Mousley (eds), *Mathematics education research: Innovation, networking and opportunity. Proceedings of the 26th Annual Conference of the Mathematics Education Research Group of Australasia* (Vol. 1, pp. 499–506). Geelong: MERGA.

Luke, C. (2000). Cyber-schooling and technological change: Multiliteracies in New Times. In B. Cope and M. Kalantzis (eds), *Multiliteracies: Literacy learning and the design of social futures* (pp. 244–66). Melbourne: Macmillan.

Newman, M. A. (1983). *Strategies for diagnosis and remediation*. Sydney: Harcourt, Brace & Jovanovich.

Sullivan, P., Zevenbergen, R. and Mousley, J. (2002). Contexts in mathematics teaching: Snakes or ladders? In M.O.J. Thomas (ed.), *Mathematics education in the South Pacific: Proceedings of the 25th Annual Conference of the Mathematics Education Research Group of Australasia* (Vol. 2, pp. 649–56). Auckland: MERGA.

Unsworth, L. (2002). Changing dimensions of school literacy. *The Australian Journal of Language and Literacy*, 25(1), 11–26.

Verschaffel, L. and de Corte, E. (1997). Word problems: A vehicle for promoting authentic mathematical understanding and problem solving in the primary school? In P. Bryant (ed.), *Learning and teaching mathematics: An international perspective* (pp. 69–97). New York: Psychology Press.

West, P. (2000). Introduction to spreadsheets for Grade 3. *COM 3 Journal*, 23(¾), 27–9.

Zevenbergen, R., Hyde, M. and Power, D. (2001). Language, arithmetic word problems and deaf students: Linguistic strategies used by deaf students to solve tasks. *Mathematics Education Research Journal*, 13(1), 204–18.

CHAPTER 4

Diversity and equity

Students in a diverse setting

Students enter mathematics classrooms with a wide range of background knowledge, experiences and dispositions. These differences—which arise both before they come to school and while they are in school—create different orientations and learning experiences. For some students, their experiences will help them in their learning of mathematics, while for others considerable input will be needed to support their learning. How teachers work with this diversity is informed by their own ideologies of how differences come to exist. Some teachers will see the differences as something biological or innate in the student, whereas others will see such differences as something constructed and reified through school practices. These two extreme positions represent the nature/nurture poles of one of the most enduring and perennial debates in education.

Before continuing this chapter, we'd like you to write some answers to the questions below. Retain them for reflection after reading about the practices of the mathematics curriculum found in most classrooms.

- Why do disproportionate numbers of indigenous students (Aboriginal, Torres Strait Islander, Maori, First Nation, Inuit) fail mathematics in comparison with their non-indigenous peers?
- Why are students from working-class backgrounds more likely to fail mathematics than their middle-class peers?

- Why are students who live in remote or rural settings more at risk of failing or performing poorly at school mathematics than their urban peers?
- Why are some students whose first language is not English more at risk of performing poorly in school mathematics than their peers who speak English?
- Why is it that when these variables are combined, the risk of failing is even higher?

Equity and equality

When considering how one deals with diverse classrooms, two different positions exist. Some teachers advocate an approach whereby each child is treated the same, as they regard this as fair. Others advocate an approach which recognises that differences are evident at the start of schooling and thus, in an attempt to redress these differences, that students need to be treated differently. When consideration is given to the disparate outcomes for an equity target group—for example, indigenous students—different ideologies come into play.

● Equality view

- *Assumption 1:* It is important that all students be given a fair chance at succeeding so all students should be given the same opportunities. It is up to the student to choose which options they want. If they want to succeed, the opportunities are provided for them.
- *Assumption 2:* Discourses of equality subscribe to principles of fairness and choice.

● Equity view

- *Assumption 1:* Disadvantaged students are more likely to perform poorly at school due to different home circumstances and the practices of school mathematics not aligning with the knowledge, skills and dispositions the students bring to school.
- *Assumption 2:* If disadvantaged students are to succeed in mathematics, different opportunities need to be made available to

them in order to redress the differences in school and non-school experiences.

- *Assumption 3:* Discourses of equity subscribe to principles of justice and difference.

In considering these two very different views, it becomes clear that if students enter school with very different learning experiences—some of which are recognised in school while others are not—their chances of success are different. In the case of white, middle-class students whose experiences mean that they enter the mathematics classroom with a knowledge of colours, shapes, recognising and writing numerals, the chances of success in the first year of schooling are greatly enhanced compared with their peers who enter the first year of mathematics without these experiences. This raises the issue of what teachers then do with these two very different groups of students. In the case of the equality teacher, there is every likelihood that this teacher will follow set curriculum guidelines for Year 1 and teach this content, making slight adjustments to support the learning of the disadvantaged students. The equity teacher is more likely to adopt a teaching practice whereby the background knowledge of the students becomes the centre of the curriculum and learning experiences are based around those experiences so that learning is built from there. The school curriculum will be adjusted to suit the needs of the students but with every expectation that the students will learn—that is, the curriculum will not be watered down.

Cognitive models of difference

Cognitive models of learning, such as constructivism, recognise the ways in which different experiences orientate students towards constructing different meanings from classroom work. For example, the student who enters the classroom with an understanding of a square having more than three sides will construct very different meanings from their interactions with the teaching process than the peer who enters the classroom with an understanding of a square having four sides; this will be different again from the student who enters the classroom with an understanding of a square having four equal sides. For a considerable time in the early years, all three meanings will work well. Only when mismatches arise in students' experiences with rectangles and squares will such knowledge become problematic for them. Constructivism allows teachers to see how students construct their own meanings so that when students enter the classroom their pre-existing knowledges and experiences are framed within cognitive terms.

A purely cognitive approach can engender a victim-blaming model whereby the students who fail or perform poorly in mathematics are seen as problematic. Terms such as low ability, low motivation, cognitively inferior, delayed growth, and so on are often used to explain why some students do not progress as expected. By placing the blame on the student (or their families or peer groups), teachers believe they can do little to change learning outcomes.

Social models of difference

Rather than see students' constructions of mathematical ideas and processes as purely cognitive, social and cultural models seek to understand diverse backgrounds and the experiences brought about through those backgrounds, and the ways in which these will impact on students' construction of meanings. In viewing the mathematics curriculum as a construction of social knowledge, social models of difference see the curriculum as a representation of some knowledges and the exclusion of others. Mathematics education represents some cultural values and denies others. Under these models, when teaching practices are seen as problematic and contributing to the failure of many students, teachers are able to change their practices in order to enhance learning outcomes for all students. In order to change their practice, teachers need to recognise differences in students, their backgrounds and their orientations towards schooling and mathematics. Rather than such differences being viewed as deficits, they should be seen simply as differences. Social, cultural, gendered, linguistic and geographic differences are critical factors when it comes to success in school.

In the cognitive model, the focus is on developing cognition. Social models extend this position, at the same time recognising the social factors that impact on success in mathematics. Social factors include those brought to the learning environment by the students and those of the learning environment itself. Learning mathematics is a social activity, and as such has particular unspoken rules. The rules of the mathematics classroom will be more accessible for some students than for others. Teachers need to be cognisant of these rules in order that all students are able to participate in classroom practice.

Apart from the cohort of students with intellectual disabilities, students most at risk of failing mathematics (or under-performing) are those who come from working-class (or low socioeconomic status) backgrounds, students from indigenous cultures, students whose first language is not English, students living in remote or rural/isolated areas, students with disabilities, and particular gender groups (this is closely tied to the previous variables). Students coming from these backgrounds are disproportionately represented in the lower percentiles of most test measures. This is not to say that all students from such backgrounds will perform poorly, only that they are disproportionately represented in the lower ends of measures. This chapter discusses ways in which the mathematics education is implicated in the production of such outcomes. By better understanding the hidden biases and assumptions built into the teaching of mathematics, teachers can alter their practices to cater for the diversity of students in the classroom.

Teachers' beliefs

Teachers' beliefs about learners and learning have a powerful impact on learning outcomes. The power of teachers' beliefs is profoundly evident in the study *Pygmalion in the Classroom* (Rosenthal and Jacobsen, 1969) in which students were randomly assigned rankings and placed into a new classroom. The teacher believed that the marks were the students' academic rankings. Even though the rankings were totally random, at the end of the teaching year students confirmed the ranking originally given. This study showed how the teacher's beliefs about the students created different expectations and learning environments, which in turn produced the expected outcomes. The implications of this work are that when a teacher believes a student to be able or not able, then learning experiences will be provided that will confirm this expectation. More recently, a large-scale study in the United Kingdom (Askew et al., 1997) showed that the most effective teachers of numeracy were those who believed that all students could learn mathematics.

● The myth of ability

Perhaps one of the most pervasive practices in mathematics—which is not so evident in any other curriculum area—is founded on a belief in ability. Teachers of mathematics are more likely than those in any other curriculum area to support the notion of ability grouping. At the end of a teaching year, unsurprisingly, results tend to reinforce the ability groupings made, leading teachers to believe that such groupings were appropriate. In a number of recent studies of ability grouping in mathematics in the United Kingdom, the United States and Australia (Boaler, 1997; Boaler et al., 2000; Zevenbergen, 2001), some consistent findings emerged:

- Students in high streams were most advantaged by having good teachers, being exposed to extensive curriculum and entering examinations with a good knowledge, which allowed them to succeed in the exams. In the Australian study (Zevenbergen, 2001), students reported that they would enter the exam already with a pass grade and work towards higher grades. In the UK study (Boaler, 1997), students in the high stream also reported that the pacing of lessons was very quick, as teachers expected that they were the 'bright students' and could cope with the work. The ethos of these classrooms was work-orientated.
- Students in the low streams were exposed to low levels of curriculum; were in classes with considerable management issues; covered minimal content; entered exams with low levels of knowledge and could only gain low marks.

These learning environments produced very different orientations towards mathematics as well as different achievement levels. The high-stream students were more likely to report a positive attitude towards mathematics and an intention to continue with study in the area. Students in the low streams reported a frustration with mathematics, an inability to move out of the low streams due to the organisation (rather than their desire) and a dislike for the subject. These different learning outcomes can be viewed as a consequence of learning environments based on a belief in ability. The ultimate outcomes reinforced teachers' beliefs about the students' abilities in mathematics—in fact, questions need to be posed about the teaching practices and their role in the construction of 'ability'.

Similar studies have shown that teachers hold powerful beliefs about students' backgrounds as a factor in determining success. In a study of teachers' beliefs about why students from socially disadvantaged backgrounds tended to under-perform in school mathematics, Zevenbergen (2003) found that the teachers generally blamed personal attributes of the student or family for their lack of success. In this study, teachers cited their experiences with minimal resources, namely calculators. In one case, the teacher described the students as having no respect for school property or themselves. When a structured borrowing system was put in place, teachers believed that students would damage or steal the class set of calculators. In another school, teachers saw students as needing to be respected and valued as people; here the students could borrow the calculators on an honour system. In this school, all calculators were returned undamaged at the end of the year, despite no policing of equipment. These two experiences starkly highlight the ways in which teachers' beliefs about students can impact significantly on practice.

Home–school differences

Students from socially disadvantaged backgrounds have been found to be exposed to very different literacy practices than their middle-class peers. In a study of preschool interactions, Walkerdine and Lucey (1989) found that mothers interacted very differently with their children depending on their social background. Similar generic studies conducted by Heath (1989) and Delpit (1995) have shown similar patterns of interaction. Where patterns of interaction at home are language rich—as in, for example, 'Please get your yellow jumper. It is on the top shelf of the middle section of your wardrobe, just next to the pile of T-shirts', it can be seen that the language is not only very rich in the genre of school interactions, but also rich in the language of mathematics. The studies cited have shown the interactions in the home of socially disadvantaged students are more likely to be of a declarative statement depleted of adjectives—'Go and get your jumper'. Those in the first group of students are exposed to a rich spatial language—'top', 'middle', 'next to'—as well as to the language of colour, which is one of the key tools for early years sorting and classifying activities. Many teachers bemoan the deficits of disadvantaged students when they come to school unable to write their numerals and not knowing colours or basic shapes. It is not that these students have lower levels of cognitive development, it is that their family environment does not use school English. Students whose home environments expose them to the rich language that prepares them for school are better prepared for many aspects of school mathematics. It is not that they are more intelligent, only that they have been exposed to home practices that will prepare them more effectively for school than their less-advantaged peers.

Other studies have shown similar differences in specific aspects of language. Cross-cultural studies have indicated that the language used by different cohorts of students can be very different even when they speak English. In her work with young students, Walkerdine (1992) found that middle-class parents were more likely to use the terms 'more' and 'less' in their interactions with their children—'Would you like a little more or a little less meat with your dinner?'—whereas working-class parents were more likely to only use the signifier 'more'—'Would you like some more meat with your dinner?' or 'Don't you want any more?' Similar cross-cultural work has found the same differences in language use with many indigenous peoples. This subtle difference has considerable manifestations for early work in mathematics. In these strands, teachers will be asking questions such as 'Which number is 2 more than 3?' 'What is more—3 or 5?' 'What number is 5 less than 8?' 'Which column has more than this one?' when undertaking graphic work. Such questioning is integral to most early teaching of mathematics. However, when young students are exposed to these forms of language and it is not seen as problematic, potentially many of the students from disadvantaged backgrounds may be missing significant chunks of the content due to their unfamiliarity with the terms 'less' and 'more'. Rather than seeing the students as deficient in their understandings, teachers should recognise that the language of the students' backgrounds inhibits their potential to make sense of the lessons.

More subtle differences emerge when other aspects of mathematics teaching are considered. Current reforms such as the use of open-ended tasks change the culture of mathematics, but this is never made explicit to students. In studies of curriculum reform using open-ended tasks, various aspects of pedagogy were found to be problematic for young students. Lubienski (2000) found that her low socioeconomic status

(SES) students preferred working on closed tasks, whereas middle SES students reported a preference for open-ended questions. Many students have constructed a sense of mathematics as being a discipline with one correct answer, but when this is changed (as in the case of open-ended tasks), teachers need to be cognisant of the need to make the criteria explicit to students. Similarly, when using group work or setting students to tasks such as 'discuss the problem in small groups', teachers need to explain the rules and expectations (Sullivan et al., 2003). While this may be seen as good classroom practice, it is particularly important for students from socially disadvantaged backgrounds, for whom the purpose of the tasks may not be as obvious as it is to their peers. Without the expectations of the task being explained, students may be at risk of not undertaking the task in a way envisaged by the teacher, due to their inability to recognise the implicit rules of group work rather than the mathematical demands.

Many of the tasks set for students have been embedded in word problems with the intention of making the task 'real'. However, the use of contexts can create difficulties. Most obviously, the immediate readability of the text can be problematic for students whose grasp of English is not strong. However, another layer of difficulty is added as there are now two competing discourses operating; the students need to identify whether the task is a school mathematics one or a real task (Cooper and Dunne, 1999). Working with students in the upper primary grades, Zevenbergen and Lerman (2001) reported that in order to be able to respond correctly, students need to identify the task as mathematical, select the elements that will enable a correct response to be generated, then respond in a format that is seen to be appropriate in that context.

In a large-scale study of similar questions posed on a testing scheme in the United Kingdom, Cooper and Dunne (1999) found that working-class students were more likely than their middle-class peers to respond in ways that suggested they interpreted the questions as real problems rather than school mathematics and thereby to offer incorrect responses. Middle-class students were more likely to see the problem as a school maths task and answer in that mode. Cooper and Dunne suggest that middle-class students are more able to code-switch between the contexts of home and school, whereas working-class students are more likely to work within everyday contexts and thereby misinterpret the questions and ultimately offer incorrect responses. When follow-up interviews were undertaken, the students were able to calculate the answers correctly once they realised that their interpretation of the question was incorrect, thus suggesting that the problem is not so much mathematics but rather the registers within which the questions are posed and that the contexts are not made explicit to students.

Knowledge and world-views

One of the reasons for differences in school performance is the world-views with which students enter school—that is, their particular ways of seeing and viewing the world. In most Western societies, considerable emphasis is placed on quantifying things. Young children are indoctrinated with these world-views from the time they enter the world. How old is your friend? How many brothers and sisters does she have? What date is your birthday? Everyday events are talked about in numbers so the idea of numbers, counting and measuring is an integral component of the worlds in which most students live.

For those who live in a quantifying society, many of the ideas used in mathematics may be seen as unproblematic. Take the teaching of percentages. Often it is taught in the contexts of profits, mark-ups or discounts as if these are part of the natural order. Making a profit—as represented in the idea of mark-ups—is legitimised through the teaching of percentages (Gill, 1988). It is taken for granted. But is this so for everyone? In egalitarian and many agrarian cultures, there is no notion of profit or mark-up and the idea of profiting from your peers is the antithesis of community.

The cultural knowledges with which people are familiar can be illustrated by the fish in water analogy. When in the water, the fish assumes the world is as it should be; however, when the fish is removed from the water, the world is very strange. World-views work in a similar way. Even within Western cultures, things are done differently in different countries—how to catch a bus, how to queue, how to pay for items. The differences become even greater when the cultures are even more diverse.

The base 10 counting system is commonly used in Western cultures— to the point where it is seen by most members of the public as the 'natural' way to count. Yet, while school mathematics shows other counting systems based on different number bases, other 'natural' counting systems exist in other cultures. For example, in his extensive study in Papua New Guinea, Lean (1996) recorded hundreds of different counting systems, many of them based on body parts. It is argued that more complex counting systems arise from the need for counting large numbers of objects. In many nomadic cultures, there is no need for large numbers as the society hunts and gathers with little trading. When cultures begin to trade, a need for a more complex number system arises. In Western society, the demise of the Roman numeral system resulted from its complexity when dealing with large numbers; the Hindu-Arabic system used in most countries allows for large numbers and undertaking work with those numbers.

Measuring time is a construct unique to quantifying societies, represented by phrases such as 'time is money' and 'saving time'. It is a construct that has arisen from a need to communicate about time. In order for Western society to work well, points in time and the passage of time need to be measured. Often such constructs are taken for granted, as if they were natural phenomena, whereas in fact they are social and historical artefacts (Poole, 1998). Even within Western societies, there are differences—for example, consider the seasons. In some countries in the Southern Hemisphere, spring and autumn begin on the first days of September and March; in other countries, they begin on the equinox (when day and night are of equal length), which is around the 22nd or 23rd of those months. This indicates that the timing of the seasons is merely a convention.

In some indigenous cultures, seasonal attributes are considered from a qualitative perspective. In her work with Australian Aborigines, Harris (1990) notes that the equivalent of a calendar was discussed in terms of qualitative features—the weather, the flora and fauna—so that in any one year a season would be as long as the qualities were apparent. Rather than see summer as being three months in the period from December to February, she suggests that summer would be as long as particular qualities remained apparent.

The subtle power of how people come to see the world in particular ways can be seen when teaching area. In this section of the curriculum, common examples relate to students undertaking calculations centred on areas of various shapes. Sometimes decisions need to be made about choices, which inevitably are made on attributes of size. For example: 'A farmer is given the choice of two blocks of land—one is a rectangle of 7 km $\times$ 2 km, the other one is a square of 4 km. Which one would he choose?' The expectation would be that the students would calculate the area and choose the bigger block. Students with a knowledge of farming might also decide that the square would be better because it required less fencing material and therefore was an even better proposition. When similar problems were posed to students in Papua New Guinea, Bishop (1988) found that the students' reactions were not based on quantification; rather they posed more questions about the quality of the land—how fertile it was, the slope (too much would create erosion), and so on. These examples suggest that students' world-views may position them very differently in terms of how they will interpret the problems that are posed, and how they are generally orientated towards working mathematically.

How people come to see their worlds impacts significantly on how they think and perform mathematically. It is not that there are differences in innate ability, but rather differences in cultural knowledges. For some students, it is not simply a case of learning concepts and processes that extend their familiar knowledges but something much more complex. This applies not just to students from very diverse cultures, as in the cases cited here, but also to students whose gender, social class and geographical location (i.e. rural/urban) are different.

Implications for school mathematics

When students from diverse social and cultural groups come to the mathematics classroom, they bring with them a wealth of knowledge and experiences. Some of these experiences will be acknowledged and others will be denied. The counting systems of Western cultures are seen as legitimate mathematics, and where indigenous counting systems are taught (which is infrequently), they are seen as novelties rather than as legitimate mathematical activities.

When a teacher becomes aware of the differences in knowledges and how these relate to school mathematics, there is a significant opportunity to adjust practice in ways that can allow students from diverse backgrounds to come to learn and understand mathematics.

For example, when teaching ratio, one of the most common teaching examples is for students to calculate the best buys of commonly purchased items. In attempting to be inclusive, a teacher may assume that students from wealthy backgrounds would purchase expensive items whereas students from economically disadvantaged backgrounds would purchase cheaper items, and so use economically differentiated items for the worked examples. However, when considering the backgrounds of disadvantaged

Indigenous students working with a number board

students, it is not uncommon to find that they engage in fishing, as this is a cheap activity and one that can support the family. It is also one rich in ratio knowledge—the hook needed for small fish such as whiting is very different from the hook needed for large fish such as mackerel, for example—which is seldom used in mathematics examples. This knowledge would be familiar to many disadvantaged students and would lead to success in understanding ratio, as it has relevance and purpose. It also legitimates the knowledges of the students so that they feel they are a valued part of the mathematics culture.

In a large study aiming to identify and break down many of the barriers to learning mathematics for socially, culturally and linguistically diverse students, Sullivan and colleagues (Sullivan et al., 2001) identified a number of features that teachers need to adopt:

1. Be explicit about expectations for work and assessment—do not expect students to second-guess what is expected of them.
2. When using contexts for mathematics problems, ensure that the contexts are relevant and sensitive to the students' needs.
3. When undertaking mathematical activities that have elements of language to them, work through the tasks so that students understand what the language is asking of them—this may mean writing new words on the board (with meanings); undertaking language-based mathematics lessons (just as in English classes—spelling, textual construction and deconstruction, etc.).

4. Explain the pedagogies being used—why group work is being used, the roles of people in groups, the expectations of the various group members, the anticipated outcomes for the group, and the reasons and benefits of such approaches to the type of lesson being conducted.
5. When introducing new approaches to teaching mathematics—group work, open-ended tasks, investigations—spend time explaining or modelling to students the nuances of the approaches so they are not expected to second-guess what is expected from them. This is particularly important when the approach being used is very different from the usual classroom practices.
6. Believe that the students can learn!

Perhaps the most important factor in attaining success for students who are more at risk is for teachers to have expectations of learners. Many educators working with Aboriginal students, for example, strongly advocate that teachers must have high expectations of their students. This will ensure that the teacher provides a high-quality curriculum with strong scaffolding to enable students to learn. When teachers work with deficit thinking, they are more likely to offer a low-level and poorly stimulated learning environment that holds back the progress of students.

■ REVIEW QUESTIONS

4.1 In what ways does the mathematics curriculum exclude some students from participating fully in classroom activities?

4.2 Imagine you are sent to a remote area where there are high numbers of socially and/or culturally diverse students. What strategies would you put in place that might support students' learning in mathematics?

4.3 In what ways could you adjust the curriculum so that you could include (in a legitimate and appropriate way) different cultural knowledges that have a mathematical orientation to them?

4.4 Reflect on what you learnt about language and mathematics, and apply it to what you have learnt from this chapter. How are mathematics, language and diversity implicated in the construction of social and cultural differences?

Further reading

Becker, J.P., Silver, E.A., Kantowski, M.G., Travers, K.J. and Wilson, J. W. (1990). Some observations of mathematics teaching in Japanese elementary and junior high schools. *Arithmetic Teacher*, 38(2), 12-21.

Hessler, G. L. (2001). Who is really learning disabled? In B. Sornson (ed). *Preventing early learning failure* (pp. 21-36). Alexandria, VA: Association for Supervision and Curriculum Development.

Hiebert, J. et al (1997). Equity and accessibility. In J. Hiebert et al., *Making sense: Teaching and learning mathematics with understanding* (pp. 65-74). Portsmouth, NH: Heinemann.

Koshy, V. (2001). Identifying mathematically promising students. In V. Koshy, *Teaching mathematics to able students* (pp. 19-28). London: David Fulton

Milton, M. (2001). Who helps children who have learning difficulties in numeracy? *Australian Primary Mathematics Classroom*, 6(3), 6-10.

Montis, K.K. (2000). Language development and concept flexibility in dyscalculia: A case study. *Journal of Research in Mathematics Education*, 31, 541-56.

Robert, M. (2002). Problem solving and at-risk students: Making 'mathematics for all' a classroom reality. *Teaching Children Mathematics*, 8(5), 290-5.

Sun,W. and Zhang, J. (2001). Teaching addition and subtraction facts: A Chinese perspective. *Teaching Children Mathematics*, 8(1), 28-31.

Weaver, L. R. and Gaines, C. (1999). What to do when they don't speak English: Teaching mathematics to English-language learners in the early childhood classroom. In J.V. Copley, *Mathematics in the early years* (pp. 198-204). Reston, VA: National Council of Teachers of Mathematics.

Zazlavsky, C. (2001). Developing number sense: What can other cultures tell us? *Teaching Children Mathematics*, 7(6), 312-19.

References

Askew, M., Brown, M., Rhodes,V., Johnson, D. and Wiliam, D. (1997). *Effective teachers of numeracy: report of a study carried out for the Teacher Training Authority 1995-96.* London: School of Education, King's College.

Bishop, A.J. (1988). *Mathematical enculturation: A cultural perspective on mathematics education.* Dordrecht: KluwerAcademic.

Boaler, J. (1997). *Experiencing school mathematics: Teaching styles, sex and setting.* Buckingham: Open University Press.

Boaler, J.,Wiliam, D. and Zevenbergen, R. (2000). The construction of identity in secondary mathematics education. In M. Santos (ed.), *Proceedings of Second International Mathematics Education and Society Conference* (pp. 192-202). Lisbon: Universidade de Lisboa.

Cooper, B. and Dunne, M. (1999). *Assessing children's mathematical knowledge: Social class, sex and problem solving.* London: Open University Press.

Delpit, L. (1995). *Other people's children: Cultural conflict in the classroom.* New York: New Press.

Gill, D. (1988). Politics of percent. In D. Pimm (ed.), *Mathematics, teachers and children: A reader* (pp. 122–5). London: Hodder & Stoughton.

Harris, P. (1990). *Mathematics in a cultural context: Aboriginal perspectives on space, time and money.* Geelong: Deakin University Press.

Heath, S.B. (1989). *Ways with words: Language, life and work in communities and classrooms* (2nd ed.). Cambridge: University of Cambridge.

Lean, G.A. (1996). Counting systems of Papua New Guinea and Oceania. Unpublished doctoral thesis. Lae: Papua New Guinea University of Technology.

Lubienski, S.T. (2000). Problem solving as a means towards mathematics for all: An exploratory look through a class lens. *Journal for Research in Mathematics Education,* 31(4), 454–82.

Poole, R. (1998). Calendars, calibration and culture. *Mathematics in school,* 27(5), 14–16.

Rosenthal, R. and Jacobsen, L. (1969). *Pygmalion in the classroom.* New York: Rinehart & Winston.

Sullivan, P., Mousley, J. and Zevenbergen, R. (2001). *Overcoming barriers to mathematics learning: Advice to teachers.* Bundoora: La Trobe University.

Sullivan, P., Mousley, J., Zevenbergen, R. and Turner Harrison, R. (2003). Being explicit about aspects of mathematics pedagogy. In N. Pateman, B. Dougherty and J. Zillox (eds), *Proceedings of the 28th Annual Conference for the International Conference of the Psychology of Mathematics Education 27* (Vol. 4, pp. 267–74). Hawaii: PME.

Walkerdine, V. (1992). Reasoning in a post-modern age. Unpublished paper presented to the 5th International Conference on Thinking, Townsville.

Walkerdine, V. and Lucey, H. (1989). *Democracy in the kitchen: Regulating mothers and socialising daughters.* London: Virago.

Zevenbergen, R. (2001). Is streaming an equitable practice? Students' experiences of streaming in the middle years of schooling. In M. Mitchelmore (ed.), *Numeracy and beyond: Proceedings of the 24th Annual Conference of the Mathematics Education Research Group of Australasia* (Vol. 2, pp. 563–70). Sydney: MERGA.

——(2003). Teachers' beliefs about teaching mathematics to students from socially disadvantaged backgrounds: Implications for social justice. In L. Burton (ed.), *Which way social justice in mathematics education?* Westport, CN: Praeger.

Zevenbergen, R. and Lerman, S. (2001). Communicative competence in school mathematics: On being able to do school mathematics. In M. Mitchelmore (ed.), *Numeracy and beyond: Proceedings of the 24th Annual Conference of the Mathematics Education Research Group of Australasia* (Vol. 2, pp. 571–8). Sydney: MERGA.

CHAPTER 5

Planning for teaching

Planning is an integral component of the teaching and learning process.

In order to develop a quality learning environment, the teacher needs to plan for short-, medium- and long-term goals. Without good planning, teaching risks becoming ad hoc and thus threatening the gains that can be made by students. Three key questions underpin effective teaching in mathematics:

1. What am I going to teach?
2. How will I teach it?
3. How will I know whether I have been successful in teaching it?

When a teacher is able to provide answers to these questions, the task of teaching becomes somewhat simpler.

In contemporary education contexts where there is increased accountability, teachers are held responsible for providing learning for their students. This focus represents a substantive shift from the recent past, where the focus was more on teaching and the emphasis was on providing activities that could engage students. Often the students could be seen to be doing 'busy work', but there was little intellectual quality to these activities—or, if there was, it was not made explicit to the students. More recently, the emphasis has shifted to student learning and high expectations on teachers to provide learning environments that will produce the desired learning outcomes. Current reforms in many places recognise the importance of focused, high-quality learning experiences. In some places, this has resulted in governments providing guidelines to teachers on how to prepare quality lessons and quality discussions within those lessons. More than ever before, teaching in contemporary times demands that teachers plan for their students' learning.

Planning

In planning short- through to long-term teaching episodes, salient and overt factors necessary for successful teaching of mathematics impact on the teaching and assessment inherent in the planning process. Two factors impact on planning:

1. *The individual's subjective orientation towards teaching.* This includes beliefs about how students best learn mathematics. If the teacher subscribes predominantly to a particular view of learning, this is likely to be reflected in their planning (including teaching and assessing).
2. *The objective or structural aspects of teaching mathematics.* This relates to aspects of mathematics education that may be perceived as being beyond the control of the teacher. Such things include set curriculum documents or schemes, and testing regimes.

● Levels of planning

Three key levels of planning guide teachers' work in mathematics:

1. *Long term*—where the teacher may develop plans for the full year's work. These can come from, and may be guided by, school policy documents, government education documents and materials from commercial publishers. Teachers may use all these more fully developed products to greater or lesser degrees, depending on how well they link in with the school and the individual classrooms.
2. *Medium term*—where the teacher may plan for an extended period of time such as a school term or a period of time within a term. Planning at this level involves units of work that may be thematic or concept-based.

3. *Short term*—where the teacher plans for a single lesson or a short unit of time such as a week. This can be undertaken by pre-service teachers as they embark on the principles of planning and the minutiae of organising for effective learning and teaching.

Planning can be undertaken at the level of the individual teacher, across teams and/or across the school. Much of this planning may be influenced by school policies.

Teacher display in a classroom showing class units of work

Why plan?

For experienced teachers, early planning becomes an internalised process, and is often barely visible. For novice teachers, the early planning of lessons is not an end-product, but rather the introduction to the planning process—the development of planning principles. Planning is an essential part of good teaching, as it has the following purposes:

- It ensures that teachers have deliberate intentions of what they want to achieve and in what timeframe. Teachers need to have clear goals about their teaching. Being focused on what is intended over a given period of time allows teaching to remain focused on particular outcomes.
- It avoids repetition. The mathematics curriculum is a very crowded body of knowledge, so teachers need to ensure that they do not repeat material unnecessarily. At the same time, they also need to allow adequate time for revision of work when this is required.
- It ensures that all content is covered. Progression through the primary school demands that students cover the content nominated. Planning allows teachers to see the big picture of their year's work so that they can ensure that appropriate amounts of time are spent on topics in order to ensure that all key content is addressed.

- It ensures that content is covered in an ordered and appropriate time-frame. Seeing the big picture of a year's work, or a smaller amount of work, allows teachers to recognise the best order in which to deliver content, and the most effective way to achieve the breadth of what is stated in curriculum documents.
- It allows teachers to cater for the diversity within a classroom.
- It ensures that the lesson/unit will be motivating for students.
- It creates a sense of confidence in what is going to be achieved.

Planning: What do I want to teach?

● Identify the intended learning for students

Central to quality learning is knowing why particular experiences are beneficial for students' mathematics learning. Consider a task in which brightly coloured paper clips are placed on various items and students are required to count and record the number of clips used. This activity, which is loosely for the purpose of bringing about the idea of informal units of length, might keep students busy—but the real point of the activity must be questioned. Is it just a counting and recording activity since the students are going to be introduced to various forms of tables? Is it about measurement? If so, measurement of what? Length? Perimeter? Using informal measurements? Is it a situation where students are expected to problematise the use of informal measurements in preparation for the use of formal units of measure? With so many questions raised, the activity has considerable potential for diversion, resulting in lack of focus. Making connections is a critical aspect of teaching, but if there is no sense about the purpose of the lesson, then it will be difficult to make coherent connections for the students.

Depending on the teaching episode, there is a range of potential foci for a lesson. Not all teaching episodes will focus on mathematical knowledge alone; they may consider other aspects of mathematical activity, including:

- conceptual—where the knowledge of mathematics is considered (e.g. fractions, subtraction)

- processes—where the 'how to' is the focus (e.g. problem-solving, inferring, using technology); this can be related to thinking and working mathematically
- affective—where students' feelings are considered (e.g. perseverance, enjoyment, motivation)
- social—where social dynamics are considered (e.g. group work, collaboration)
- language—specific aspects of mathematical language or modes of communication.

● Specific learning outcomes

Once the main outcome for a lesson or unit is identified, syllabus documents provide extra guidance through suggesting indicators of learning. Within the new approaches to curriculum development where learning outcomes form the basis of the reform, support documents have broken down these outcomes into smaller manageable units that are usable (and assessable) by teachers. These can be identified and used for planning. As various systems have adopted different formats for their syllabus documents, teachers need to become familiar with those adopted in their own locality.

● The context

Teachers need to consider the specific context within which they will teach. This will vary from classroom to classroom, school to school and year to year. It is important to consider the interests, needs and background knowledge of the students in any given classroom. This is particularly the case for diagnostic work and extension.

● Background knowledge of students

When planning any learning, consideration of what students already know is central to decision-making. Considerable time can be wasted if students continually repeat content they already know. Similarly, planning around their interests is important. Good planning involves recognising what students bring to the learning situation.

● The learning continuum

Identify the concepts that come before and after a particular activity. By knowing what is the prerequisite knowledge and follow-up knowledge, the teacher is able to cater for the diversity in the classroom as well as being prepared for the unexpected. Quite often, students may not have remembered earlier work, or may be learning beyond what has been intended. By knowing what would extend the focus of the teaching episode or knowing where there may be potential errors or gaps in learning, the teacher is better able to adjust the lesson to suit the students' current levels of understanding. For any lesson or teaching episode, teachers often need to plan for at least three different levels of achievement to ensure that all students can engage in the teaching episode through planning revision activities, consolidation activities and extension activities (see diagram).

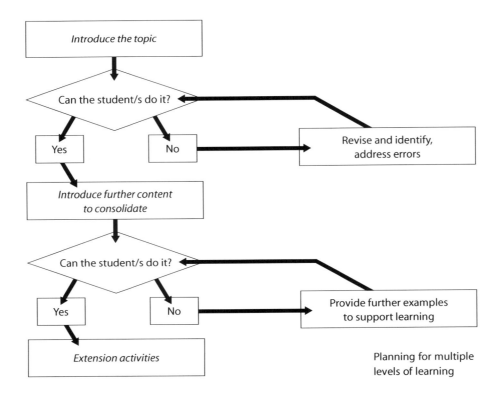

Planning for multiple levels of learning

● Common misconceptions

A knowledge of the common errors and misconceptions that occur in learning particular concepts is useful when preparing learning experiences. For example, a common error that students make when learning subtraction with trading (or decomposition) is to take the smaller number away from the larger number, regardless of its position. This type of action may be interpreted by the novice teacher as 'the students do not really understand the practice of subtraction'. Experienced teachers anticipate such errors, and planning to avoid the development of such errors is commonplace. Often described as 'error patterns', these mistakes can be due to teaching practices. In this example, the error can be due to the students internalising a process of 'taking the little number from the big number', which has worked well in their experiences up to this point, so that when they encounter this sum, they implement their construct. The misinterpretation of the process can be due to the way in which the teaching leading up to this point has been undertaken.

● Language

Mathematics is recognised as being a very particular language (see Chapter 3). Indeed, in some countries the study of mathematics is based in the language faculty. There are many words that are specific to mathematics (numerator, pi), whereas others have specific meanings in mathematics that are different from other contexts (such as mass, ruler, base). Identifying the words that will be used in a lesson or unit is very important, as they can prove to be another source of confusion for students.

● Resources and ICTs

Identifying the resources and ICTs (Information Communication Technology) needed for a lesson or unit is important. When equipment is missing—such as protractors, thermometers or the overhead projector—teaching potential is diminished as the lesson is altered or valuable teaching time is lost hunting for the missing resources. As the use of more advanced forms of technology (such as computers, calculators and graphic calculators) permeates the teaching of mathematics, their use and application need to be included in planning for learning. Learning through technology is an important and growing area of teaching in modern society.

● Inclusion of all students

Ensuring *all* students are included in the learning is a high priority in planning. Planning must consider the students who need help, those who are likely to finish early and those whose English (or the language of instruction) may not be strong. It should also incorporate students' cultural or social experiences into the learning. At the same time, the focus must remain on the mathematics learning. It is insufficient to allow early finishers to colour in or undertake activities that are not related to mathematics. Authentic experiences related to the goals of the lesson or unit should be developed so as to include and extend all students.

● Transdisciplinary planning

As teachers' work becomes more intensified and there are greater demands to make curricula more relevant to realistic settings, the demand to integrate curricula becomes more important. When planning, it is useful to investigate other curriculum areas and make links between them. Linking data collection and analysis (statistics in the mathematics curriculum) is easily achieved with the social studies curriculum; monitoring the environment can be undertaken through science; recording body function rates can occur in health and physical education lessons. Links can be made with English when writing reports of mathematical activities; patterns can be explored through music; or time can be studied through social and environmental studies. Teachers need to make as many cross-curricular links as possible in order to meet some of the curriculum demands placed on them in modern society. As the curriculum seeks to be more relevant to the world beyond schools, links across curriculum areas are essential if students are to see that learning mathematics has purpose and relevance to the world beyond the mathematics classroom.

Teaching: How will I teach?

Having identified important pre-planning issues, it is time to consider the best means of achieving the identified goals. There should be a balance between the different approaches to teaching to cater for the diversity in a classroom as well as the different demands of the curriculum. Some aspects of mathematics lend themselves better to some approaches than others. For example, many measurement activities are suited to hands-on, materials-based learning, whereas some areas of number may be suited better to whole-class, directed learning.

● Types of lesson

Three main types of lesson exist—problem-based learning, teacher-directed learning and review–teach–practise. Each lesson form has particular strengths and weaknesses. The most common is the 'review–teach–practice' format. However, the Third International Mathematics and Science Study reported that this was a common method in many countries which scored poorly on the tests. Countries such as Japan which scored highly on the test were more likely to use problems or investigations. Such data support a common belief that the review–and–revise format may not be the best method for teaching mathematics—particularly new concepts.

Problem-based learning

This type of lesson is most beneficial when encountering new content, developing problem-solving skills, applying understandings and consolidating learning. It most aligns with the ways in which people learn, use and apply knowledge in the world beyond schools. Providing a motivation for the task (such as using newspaper clippings, video extracts or a problem relevant to students' interests) engages the students in purposeful learning. Having engaged in the task, students may work individually, in pairs or in small groups to solve the problem. It is important with this type of lesson that significant time is allowed for the synthesis of ideas and processes at the completion of the investigation. The teacher should provide a rich context in which students can evaluate each other's work in terms of the ideas, processes and content used by their peers. One strategy that works with this type of lesson is the use of problem posing (Stoyanova, 1998), which is increasingly recognised as important in terms of problem-based learning.

Direct instruction

Unlike problem-based learning, where the students are the centre of learning, direct instruction sees the teacher as central. It is useful when there is a need to teach new concepts or a particular language. In this lesson type, the teacher will assume a key role in the initial part of the lesson and then have students work on the focus later in the lesson. The final phase of the lesson usually involves a teacher-led summary of the ideas central to the lesson.

Review–Teach–Practise

Arguably the most common lesson type, this format has been found to be highly ineffective in catering for student learning. The lessons tend to focus on skill development followed by practice of that skill. The approach is evident in many textbook series, where the examples are used for practising skills. Typically, lessons are disjointed so that there is little connection or application of knowledge and skills to be learnt. This hinders students from developing a deep understanding of the purposefulness and application of mathematical ideas. This lesson type can be useful for consolidation of ideas but it should be kept to a minimum, with other lesson types (particularly the problem-based format) taking a much more dominant place in planning.

● Elements of a lesson

Teaching is broken into three main phases—for both lessons and units—each serving a very different purpose within the overall structure of teaching episodes.

Orientating phase (introduction)

The orientation phase serves three key roles—to introduce the topic, to motivate the students, and to assess whether or not students have the prerequisite knowledge to move into the lesson. With problem-based learning, this phase is used to introduce the topic, to motivate students and to launch the problem.

Enhancing phase (work or consolidation)

This phase of the teaching episode enhances and consolidates student learning. Within a problem-based format, this phase is used for students' exploration of ideas. Planned activities should cater for the diversity in the classroom, and ensure that the different needs and understandings of students are addressed. Regardless of the lesson type, this phase is used to develop the ideas or skills that are the focus of the teaching episode. Teaching strategies can vary, with the same outcome/s being achieved. It is critical that the activities that are planned for this phase link with the intended learnings. If the teacher is not able to make this link, then it is questionable whether or not the students would be able to. For example, if the activities are linked to the students understanding the concept of area, then a range of activities that allow students to 'cover an enclosed space' would be a possibility—such as covering the table with pieces of coloured paper, pasting pasta on to a piece of cardboard or placing pattern blocks over a bench top. By knowing clearly what is intended, the range of activities can be sure to address that outcome.

Synthesising phase (conclusion)

The synthesising phase serves to sum up and focus on the key learning of the lesson. It should not be a 'show and tell', but rather engage the students in rich, productive discussion centred on the intended learning/s. Current research and policy in the United Kingdom focuses very heavily on this aspect of teaching. It is now a core part of mathematics teaching in the United Kingdom that teachers must make links explicit between the purpose of the activities and the proposed learning outcomes. In order to do this effectively, the teacher must be cognisant of the links. It is insufficient to just ask students to recall what they have done. It is critical that they are able to articulate the learning in what they have done. Constructivist writings have also drawn considerable attention to this aspect of classroom dialogue (Yackel, 1991). Debate among students is healthy as it allows for clarification of misconceptions and development of new (and improved) ideas. It is important that the teacher is able to facilitate discussion in such forums.

Within a unit of work, the synthesising phase may be more developed and be ongoing, both through daily lessons and at the completion of the unit. The clarity of the purpose of the teaching episode is transparent to the students and the teacher. No longer is the pasting of pasta on a sheet an activity that keeps students busy; its explicit purpose is the development of an understanding of the area attribute.

Evaluating teaching: How will I know whether it has been a success?

There are two major forms of evaluation that teachers must undertake to determine the success of their teaching. The first is the ongoing assessment that occurs throughout a teaching episode—assessment on the run; this is a formative assessment that is used to inform the teaching processes throughout the episode. It involves posing questions, listening and responding to students, and observing students' actions, writing and behaviour throughout the lesson or unit. Experienced teachers undertake this type of assessment as a natural part of their teaching.

Assessment-on-action is a more formal aspect of evaluation. It is summative, and occurs at the end of a teaching episode. Teachers need to reflect on the lesson (or unit) and consider what was effective in the lesson, what was not so effective and what would be changed if the lesson was taught again. Self-reflection at the end of a teaching episode is crucial for future planning, to ensure that all teaching episodes are in tune with the students' interests and to ensure students' continued motivation to learn. A good habit to adopt is to write a short self-reflection on each lesson as soon as possible after teaching. This could be notes on the lesson plan itself, or a reflective journal/diary. The reflective comments should note things that worked well during the lesson, things that didn't work so well and what you could do next time if you were to teach that lesson again. The focus should be primarily on analysing your own actions and practices, rather than finding excuses for why things didn't work so well on the students' behalf. If your planning was inclusive of all learners, was well structured and scaffolded all learners, ensured students were challenged but also engaged in the activities, and your lesson took a student-centred approach, then the lesson should have been a great success.

Planning the learning environment

● Grouping

Three main forms of grouping are apparent in teaching mathematics: whole-class teaching, small-group teaching and individual teaching.

Whole-class teaching is useful when introducing new work, as it can save considerable teaching time. The teacher needs to identify which students may not have understood what was being said, and then work with those students in a small-group environment. Small groups of three to six are useful to cater for various

learning situations. Groupings can be used to cater for achievement levels, learning styles, interest groups, gender groups, and so on, and both homogenous and heterogeneous groups can be used. In general, it is often more practical for whole-class teaching to be undertaken when introducing new concepts so that the teacher's workload is consistent, while for consolidating or very differentiated work, small groups are more appropriate. When structured properly, small groups can facilitate greater dialogue among students and enhance students' understandings of concepts as well as developing social skills. Similarly, individual work can be successful for consolidation of skills where an individualised learning program is clearly needed.

● Questioning

Many teachers use questioning rather than direct instruction as the basis for their teaching. Being able to do this requires considerable skill, and enables students to be actively engaged in the teaching episode while simultaneously providing the teacher with valuable information about what they are thinking. When planning teaching episodes, consider the form of the key questions to be used. The questions should move on from closed, low-level forms to include higher-order, open-ended questions. In considering the use of questions, some researchers in science (Lemke, 1990) and mathematics classrooms (Zevenbergen and Lerman, 2001) have found that particular questions—such as single-response, yes/no questions—are used for controlling the content and flow of the lesson but have limited value in terms of intellectual engagement or substantive mathematical learning. Questions that encourage students to justify their thinking or working, propose different methods or solutions, or analyse their thinking require a much deeper level of engagement and hence more time for the formulation of responses; therefore, teachers need to practise patience when posing these questions to allow time for students to formulate responses.

Good questioning is central to good teaching (Nicol, 1999; Sullivan and Clarke, 1991; Sullivan and Lilburn, 1997). Unfortunately, it has been found that most questions posed by teachers in the mathematics classroom are lower-order questions (Wimer et al., 2001), thus greater emphasis and planning needs to be devoted to posing good questions. In the general literature on education, good questions are those that foster higher levels of cognition. Typically, such questions are based on Bloom's Taxonomy and are of the form that encourage reflection, justification and argumentation (Patterson, 1999). In this typology, a good question would be in the form 'Why is that the case?' where students are expected to explain and justify their reasoning. Other types of good question are those advocated by Sullivan and Lilburn (1997), and relate specifically to mathematics. These authors argue that good mathematics questions provoke deep learning about particular content and processes. Good mathematics questions are open questions.

Mathematics questions can be of two main forms—closed or open. A closed question provides some very focused questioning, such as 'What is the mean of 5, 4 and 3?' where there is only one answer, which students can get either correct or incorrect, and one identified method for solving the task. By contrast, open-ended questions typically do not have an obvious path of resolution and have multiple answers. An open-ended question provides for greater diversity of responses across a range of levels of understanding. Consider the question, 'If 5 is the mean, what six numbers could make up this mean?' Such a question allows students to provide a range of responses, which could include numbers that are all the same through to very large and small numbers as well as common fractions and decimals. The range of responses offered by the students allows the teacher greater access to students' deep understanding of the concept of mean, as well as catering for the range of understandings across the classroom.

● Examples of closed and open questions

Closed	Open
What is the area of a sandpit?	If a sandpit has an area of 300 cm2, with dimensions of 15 cm 3 20 cm, what might its dimensions be?
What is the sum of 7 and 8?	If the total of two numbers is 15, what might the two numbers be?

To change a closed question to an open one, a simple strategy is to look at the answer to a closed question—for example, 'the area is 300 cm^2'—and then work backwards; given that the answer is 300 cm^2, the students have to work out the dimensions. The goal of the activity (in this case, area) and the process (calculating using multiplication) are the key aspects of the task; these remain preserved and the task is inverted. There will be a number of correct and different answers to this question.

A common strategy used by many teachers is to ask three good questions for an effective lesson. Developing good questions is not a simple task, and time and effort need to be spent on their development. There should be a balance between open and closed questions, since the questions perform very different purposes in the classroom. Posing questions that demand more reflective or analytical thinking may involve less interaction across the class. To ensure participation from a wide range of students, a useful strategy can be to use the 'think–pair–share' strategy, whereby students formulate a response, share with a peer in pairs, then move to bigger groups—which could include groupings of four, eight and/or the whole class. This strategy is less threatening and allows students the time and space to work on their responses. Sharing with peers can also help to fine-tune, adjust or rethink their responses and thinking strategies.

● Materials

Teaching many aspects of mathematics involves the use of materials. It is often assumed that students will make the connections between equipment and concepts, yet research has shown that this is not always the case—indeed, when used inappropriately, equipment can create greater confusion if students are not able to see the purpose or connections between the concrete and the abstract/symbolic work (Howard and Perry, 1997; Perry and Howard, 1994). The equipment used should be chosen carefully so that it matches the intended learning. When introducing equipment to students, protocols need to be developed for how it will be used, distributed and packed away.

● Textbooks

Many primary schools opt for some mathematics scheme where students will have their own workbook. The topics to be covered in a year level may be guided by the writers of these programs rather than another authority, such as state offices or school-based documents. While there is considerable heated debate as to the value of textbooks in the primary school, there is certainly a place for quality textbooks that support teachers and students in quality learning experiences (Shield, 1998).

Textbooks are used badly in lessons when they become the lesson plan. That is, the teacher demonstrates a procedure on the board (taken from the textbook) and the students copy and then complete practice exercises from the textbook. The place for textbooks is to support a teacher's planning, rather than to replace it. When mathematics is dominated by textbook-based lessons, there is little creativity or support for learners at work. This form of teaching is often referred to as 'shallow teaching' (Vincent and Stacey, 2008) as it does not allow for the growth of deep knowledge, favouring procedural understandings rather than rich mathematical understandings linked to other aspects of mathematics. Unfortunately, many teachers adopt this approach, which effectively limits the development of richness in mathematical understandings.

Planning for substantive learning

In this chapter, the discussion has centred on various aspects of planning. When the teacher is clear about the planned outcome/s of the lesson or unit, how it will be taught and how to assess it, a coherence between the three phases of the planning (orientation, enhancing and synthesising) emerges. Knowing what is the best learning environment to achieve the highest learning possible becomes central to planning in mathematics. It is not sufficient to undertake busy work—the question of substantive learning, in concert with both the ethical and legal responsibilities of the teacher, is called into question.

Planning to engage students with deep knowledge and deep learning, as opposed to the rote learning of skills and drills, is now seen as a key reform in teaching mathematics. Too many students leave school without an understanding of mathematics—they 'do' mathematics, but do not understand mathematics. Developing practices that foster substantive learning is important to good teaching. This demands that teachers seriously consider what will be learnt and how best to organise learning. Encouraging practices where students are able to debate, contest and challenge can foster the lifelong skills needed for life in the wider community. Employing practices where students can engage with mathematical thinking and problem-solving of realistic and important problems is vital for the world beyond school.

■ REVIEW QUESTIONS

5.1 Why is planning important in mathematics?

5.2 What are some of the key considerations to be made when planning?

5.3 Devise some open-ended questions.

5.4 When planning assessment, what are the four considerations teachers make? Briefly outline what they are.

5.5 Compare and contrast the three main forms of grouping in the teaching of mathematics.

5.6 How will you know if your teaching is effective? Draw up a list of questions that will enable you to become a more reflective practitioner.

Further reading

Patterson, M. (1999). Questioning techniques. *Teacherlink*, 8(1), 14–16.

Sullivan, P. and Lilburn, P. (1997). *Open-ended maths activities: Using 'good' questions to enhance learning.* Melbourne: Oxford University Press.

References

Howard, P. and Perry, B. (1997). Manipulatives in primary education: Implications for teaching and learning. *Australian Primary Mathematics Classroom*, 2(2), 25–30.

Lemke, J. (1990). *Talking Science: Language, learning and values.* Norwood, NJ: Ablex Publishing Company.

Nicol, C. (1999). Learning to teach mathematics: Questioning, listening and responding. *Educational Studies in Mathematics*, 37, 45–66.

Patterson, M. (1999). Questioning techniques. *Teacherlink*, 8(1), 14–16.

Perry, B. & Howard, P. (1994). Manipulatives: Constraints on construction? In J. Geake (ed.), *Challenges in mathematics education: Constraints on construction* (Vol. 2, pp. 487–96). Lismore, NSW: Mathematics Education Research Group of Australasia.

Shield, M. (1998). Mathematics textbooks: Messages to students and teachers. In C. Kanes, M. Goos and E. Warren (eds), *Teaching mathematics in New Times: Proceedings of the 21st Annual Conference of the Mathematics Education Research Group of Australasia* (Vol. 2, pp. 516–23). Gold Coast, Qld: MERGA.

Stoyanova, E. (1998). Problem posing in mathematics classrooms. In A. McIntosh and N.F. Ellerton (eds), *Research in mathematics education: A contemporary perspective* (pp. 164–85). Perth: Mathematics, Science & Technology Education Centre.

Sullivan, P. and Clarke, D. (1991). *Communication in the classroom: The importance of good questioning.* Geelong: Deakin University Press.

Sullivan, P. and Lilburn, P. (1997). *Open-ended maths activities: Using 'good' questions to enhance learning.* Melbourne: Oxford University Press.

Vincent, J. and Stacey, K. (2008). Do mathematics textbooks cultivate shallow teaching? Applying the TIMSS Video Study criteria to Australian eighth-grade mathematics textbooks. *Mathematics Education Research Journal,* 20(1), 81–106.

Wimer, J. W., Ridenour, C.S., Thomas, K. and Place, A. W. (2001). Higher order teacher questioning of boys and girls in elementary mathematics classrooms. *Journal of Educational Research*, 95(2), 84–92.

Yackel, E. (1991). The role of peer questioning during class discussion in second grade mathematics. In F. Furinghetti (ed.), *Proceedings of the 15th PME conference* (Vol. 3, pp. 364–71). Assisi, Italy: International Group for the Psychology of Mathematics Education.

Zevenbergen, R. and Lerman, S. (2001). Communicative competence in school mathematics: On being able to do school mathematics. In J. Bobis, B. Perry and M. Mitchelmore (eds) *Numeracy and Beyond. Proceedings of the 24th Annual Conference of the Mathematics Education Research Group of Australasia* (pp. 571–8). Sydney: MERGA.

CHAPTER 6

Assessment

The new and innovative ways of teaching mathematics discussed in the previous chapters attempt to challenge the dominance of teacher-directed models of teaching. Similar changes are posed in this chapter, where the dominance of pencil-and-paper testing as a means of assessment is challenged. As new theories and research have influenced teaching approaches, assessment has similarly been opened up. Assessment does not exist as an entity in and of itself; it is inextricably bound to the teaching and learning process. So while this chapter is written as a stand-alone resource, assessment should be seen as integral to teaching rather than as an activity divorced from teaching and learning.

When considering assessment, three aspects need to be considered:

1. Why assess—what are the purposes of assessment?
2. How to assess—what assessment tools can be used?
3. Monitoring and reporting assessment outcomes.

Why assess?

In the past, assessment was undertaken predominantly to report to parents about their child's progress in school. Contemporary assessment practices in mathematics are multifunctional, and assessment is now used to:

- monitor student progress and document the outcomes of student learning
- support student learning and progress through the curriculum
- inform teaching practice and improve programs, and
- report to governments and funding agencies about student outcomes at state and national levels.

● Monitoring student progress

Teachers need to monitor student learning and progress continually through a variety of means. Rather than focusing just on knowledge and skills, a wider range of attributes and attitudes is now monitored. Monitoring is a continual process whereby teachers systematically collect a range of evidential pieces in order to assess student progress. The students are often involved in the monitoring process as well.

A contemporary move in education, where accountability is paramount, has seen assessment move towards more transparent processes. This involves criteria being made explicit so that students are able to see what they will be assessed against. The assessment is then marked against a rubric, where the marks to be allocated are clearly indicated. This process enables teachers to identify whether or not students are moving towards the intended learning outcomes as articulated by the teacher.

● Supporting student learning

Assessment is used as an informative process—or formative assessment—where information gained through assessment is fed back into the teaching process in order to support and enhance student learning. Assessment becomes a tool for diagnosing what is known, or unknown, and this information is then used to inform subsequent practice to move students' learning in the direction desired by the teacher. Programs such as the *First Steps in Mathematics Program* (Department of Education and Training Western Australia, 2004) or the *Early Numeracy Research Project* (Department of Education and Early Childhood Development Victoria,

2006) focus on the use of developmental or diagnostic frameworks, with students assessed and mapped against these frameworks. These frameworks help teachers to understand why students are able to do some things and not others; support teachers in realising why some students may experience difficulties in learning mathematics while others do not; and show the challenges that students may face in moving their thinking forward in particular areas or with particular concepts in terms of potential misunderstandings or misconceptions, with the goal of achieving the desired learning outcomes. These programs are also useful in helping teachers to interpret the responses made by students in light of the frameworks. This type of assessment is interventionist (Wright et al., 2000) in that it is used as the basis for subsequent action. It is also a form of scaffolding, whereby teachers identify what their students know or can do and then provide appropriate learning environments through which the students are enabled to move towards more complex understandings.

● Making informed decisions about instruction and programs

Assessment allows teachers to identify what aspects of their teaching and programs are enhancing (or hindering) student learning. Using assessment to evaluate programs or learning episodes, teachers are able to identify the strengths and weaknesses in their methods and move towards changing and improving their teaching.

● Reporting to external agencies

In the past, teachers have used assessment to report to parents about their child's progress in school. Employers have also used school assessments for staff selection. Thus schools and teachers have used assessment to report to the wider community about the successes of students in their classes. Mathematics has been a high priority within the overall curriculum, as it is a key life skill and an important discipline in most areas of work and life.

More recent trends in education, in an increasingly economically driven world, have seen teaching and schools bound by external agencies in terms of reporting. Contemporary reforms from state, provincial and national agencies have seen a greater accountability of government funding, so governments want to see clear outcomes for their spending. This applies within most Western economies, where increasingly governments are demanding documentation of student outcomes. This has resulted in considerably more testing by authorities at the systems level to document student performances.

● National testing

In recent years, most countries have moved to some form of national testing. In Australia, the testing has been focused on literacy and numeracy. Under the auspices of the Australian Curriculum, Assessment and Reporting Authority (ACARA), national curriculum, testing and reporting have now become a reality for Australian teachers. In terms of assessment, national tests are conducted in the middle of the year (around May) and are referred to as the National Assessment Program— Literacy and Numeracy (NAPLAN). While there is a lot of debate as to whether the tests are 'numeracy' or 'mathematics', the reality for teachers is that they are required to undertake testing in Years 3, 5, 7 and 9 under strict testing conditions. This is intrinsically a problem at present, due to the ways in which the years across the states and territories are structured. For example, in Queensland and Western Australia, Year 7 is the last year of primary school, whereas in the other states it is the first year of secondary school. This difference impacts on the learning and curriculum to which students are exposed. Test results are published on the ACARA site and schools are able to see how well they have performed against other schools. This makes the outcomes of these tests open to public scrutiny. This public show of information had been challenged by teacher unions and other groups, but the government has been steadfast in making these results public. Increasingly, there is pressure on schools to enhance the performance of learners, and this testing process has been the basis for government decision-making in this area. Funding is also tied to testing, causing difficulty for some schools.

While there is criticism of the value of the tests or what they are testing, they are now a reality for schools. Teachers can use the results of these tests to inform their teaching and to gauge whether or not they are improving or enhancing learning. Similarly, where there are sustained differences over time, questions need to be raised about the performance of students—in particular why there is sustained failing. Most importantly, the tests have shown that there are consistent patterns of failing among particular groups—remote Indigenous students are the most at risk in Australia. The complexities of why this occurs must be considered by teachers, communities, schools and agencies if there is to be some change to the practices in these areas. Issues of language, culture, attendance and relevance all contribute to the lack of success in these and other communities.

Models of assessment

Typically, models of teaching, learning and assessment have been seen as linear, with the teacher planning a lesson or unit, undertaking the teaching and then assessing the overall teaching episode in terms of what the students may have learnt. Such assessment may employ a range of techniques but generally follows the plan–teach–assess process.

Linear models of assessment

In contrast, models of assessment should use assessment to inform teaching so that a more cyclical approach to planning, teaching and assessing is implemented. Such models see assessment as an integral component of good teaching. Using the information gained through assessment, good teachers are able to improve their understanding of the teaching/learning process, and use this understanding to improve the learning environments for the students in their classes. Current emphasis in good teaching requires teachers to be reflective about their teaching, so incorporating what has been learnt through assessment is an important part of that process.

If what has been learnt through the assessment process is not fed back into the learning loop, then one needs to ask some serious questions about the purposes of assessment. For example, when teaching a particular concept or process in mathematics, teachers should be

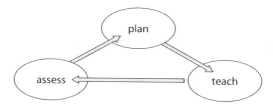

Cyclical models of assessment

assessing what students know (and don't know) and then using this information to decide where to move next in their teaching. At the simplest level, if students understand what is being taught, the teaching can move forward. If the students do not appear to understand, then learning needs to be consolidated or revised rather than moved forward. Hence assessment is an integral component of the learning loop.

Contemporary assessment practices have seen quite dramatic shifts in how assessment is undertaken and the purposes of that assessment. Some of these shifts can be seen in the table, where the move from Friday tests (usually developed by the teacher or taken from a resource book), on the basis of which students were graded using a scoring method, has been replaced by teachers using a much wider range of assessment techniques that are built into their teaching practice.

The changes in assessment practices can be seen in the report cards being used in schools. Consider the 1960s-style report card illustrated in the photo, where students were given a mark for each curriculum area and then a general comment. While the card may have been less time-consuming than those used in today's classrooms,

Examples of old and contemporary assessment practices

Old practices	Contemporary practices
Gather data periodically	Gather continuously
Assess as isolated facts or skills	Integrate assessment with teaching
Rely on tests and quizzes	Use a range of tools
Externals assess students	Students self-assess
Use a single source of evidence	Need multiple evidence
Indicate right/wrong answers	Continual feedback to students

1960s-style report card

it tells little about what the student knew, could do, needed support with, was confident in, and so on. In some cases the raw score was entered against the class average so that parents could relate to where their child performed against the class, but this did little to convey the learning of the student. However, this type of report is still seen by many parents and members of the wider community as valid assessment. They want to know students' 'scores' on tests. They want to know where the student is in relation to others in the classroom—that is, their ranking in the class—and there can still be considerable pressure on the teacher to conform to these expectations.

Modern report cards tend to be more descriptive, and reflect the strengths of the students. Often couched in positive terminology, they can, however, be criticised for failing to identify areas of weakness. More recent reporting has demanded that teachers report what the student knows or can do against curriculum requirements, but this is often couched in a language beyond the parents' comprehension.

In modern society, the mathematics teacher needs to be aware of the pressures and be able to justify contemporary thinking about assessment and reporting. Within contemporary practices, particularly using outcomes-based education, the emphasis is on documenting learning rather than rankings.

Alongside the national testing systems, there is mandatory reporting against which teachers must also report. This includes the use of rankings on an A–E scale. Government argues that this is a requirement demanded by parents. Education systems and teachers may resist this pressure, but unfortunately the pressure of government means that the schools must report using these scalings. Many schools also report using other forms of reporting alongside the mandated reporting requirements.

What to assess

When thinking about assessment in mathematics, it is important to consider what is being assessed. Past practices of testing often assessed relatively trivial aspects of mathematics—that is, knowledge and skills. More complex aspects—such as attitudes towards mathematics, problem-solving and mathematical thinking—are more difficult to assess through simple tools such as tests. Contemporary assessment in mathematics extends what should be assessed to include mathematical knowledge, using and applying mathematics and dispositions towards mathematics. These aspects of mathematics need to be incorporated into the assessment practices.

A common misperception of the testing process is that it is an objective tool for assessing students' mathematical understanding. For example, research into the language of mathematics points to some very critical points. Pimm (1991) holds that approximately 80–90 per cent of mathematical word problems can be solved using a key word strategy. Consider the two questions in the diagram.

1. My bottle holds half as much as John's bottle. If John's bottle holds 4 litres, how much does mine hold?

2. My bottle holds half as much as John's bottle. If my bottle holds 4 litres, how much does John's hold?

Posing these questions to students produces some interesting results (Zevenbergen et al., 2001). Using a key word strategy, students were likely to identify the key information as the '4 litres' and the 'half', and since half generally means 'to divide by 2', they were likely to produce an answer of 2. This works for question 1, but not for question 2, which requires a doubling process. A correct response to question 1 may be interpreted as 'the student understands and is able to undertake the mathematics within the question'. This leads to an assumption that the student is working with the ideas, knowledge and skills needed for fractions and volume. The same key word strategy does not work on question 2, and the teacher may assume that the student is having difficulties with volumes and/or fractions. In fact, there is little evidence of a mathematical strategy or knowledge being used in either working. Thus the responses to questions, and the questions themselves, need to be considered beyond what is often assumed to be the understanding of the student.

Furthering this idea, research by Ellerton and Clements (1997) on the questions posed in tests has indicated that there is potential for a gap of 28 per cent between the performance of the student and student understanding. They propose that such differences can be the result of students being able to give 'correct' responses without understanding the task, or conversely having full or partial understanding but giving an incorrect response. Extending this work, Zevenbergen (2000) analysed the questions posed on a state mathematics test for language barriers and showed how language impacted on interpretations, thus creating difficulties in the testing process. These findings challenged what was being tested—mathematics, language or cultural background—and pose serious challenges to the 'objectivity' often associated with test questions. It is not safe to assume that test items test only mathematical knowledge and skills. Rather, other aspects of learning are being applied and tested.

The limitations of pencil-and-paper testing have been well documented in terms of what is being assessed and how the evidence that is collected through the process can be severely limited and biased (Ellerton and Clements, 1997). Thus other forms of assessment may be more reliable in assessing what students know, can do and feel about mathematics. Good teachers use more than one method for assessing student learning; when various methods produce similar evidence, there is greater probability that the students understand the mathematics.

Planning assessment

Much like a researcher or detective, a teacher needs to identify what information is needed from the assessment process, how to collect the information, how to make sense of the information and how to report the results. The assessment process is just like good detective or research work—and the target of the information-gathering could be the student, the program or the teacher. The assessment process is divided into four components:

1. identifying what is going to be assessed and what will be the most appropriate tool for collecting the information
2. collecting the information

3. analysing or interpreting the information, and
4. using the results.

The ways in which the results will be used will depend on why the assessment was undertaken. For example, if the reason for assessment was to monitor student learning, then the results will be used to inform teacher action in order to create a learning environment that will promote learning. If the reason for assessment was to inform instructional decision-making, then the results will be used to determine whether or not the next lesson will consolidate learning or move on to new knowledge.

In the initial stages of planning assessment, the teacher needs to consider the actual purpose of assessment. This will inform what methods will be most appropriate for collecting the information needed. For example, if the teacher needs information about students' mis/conceptions about fractions, it becomes necessary to think about the best method for collecting such information—observing when working through tasks, listening to conversations with peers about the tasks or a worksheet could all be useful tools for assessment.

Having collected the information, it is then necessary to think about how to make sense of what has been collected. This involves the data relating to individual students, to a group and to the whole class, as well as the overall responses made by the students. Analysing one student's responses can be useful for determining what strengths or problems that student may have with fractions. Collectively, the data may show that the whole class is making a common mistake—that there is something else other than an individual misunderstanding—so something in the teaching process may need to be considered.

Once the data have been interpreted, consideration of how the assessment results will be used must be undertaken. Individual results may be used to inform intervention strategies to support development of a student's knowledge. This process is often to referred to as diagnostic teaching. Class results may be used to plan whole-class teaching or programs so that revision of previous learnings may be implemented—for example, more work with fraction kits may be undertaken by the whole class.

The information-gathering processes used are an important consideration. Some assessment tools are better than others for the information being sought. Worksheets and tests may produce some information, but it may not be detailed enough to provide the insights that are needed; in this case, discussions with the student may be more appropriate. If students are experiencing difficulties with the concept of area, for example, observing them as they are working with covering activities can facilitate the teacher's understanding of their working processes. The ways in which students work with the blocks or talk with peers can provide valuable insights into their thinking, which can further be supplemented by asking questions as they work through their tasks.

When undertaking assessment, through whatever means have been decided, it is equally important to consider the criteria that will be seen as evidence for the desired learning. Observing students for what would constitute evidence of learning (or confusion) is a key consideration.

Making assessment criteria explicit allows the teacher to consider the expectations that should emerge from the teaching, as well as the common errors that might occur and the misconceptions that can develop. Being cognisant of what can and should be expected from a given teaching episode (a lesson through to units), teachers are able to assess more effectively. The teacher notes in the diagram indicate a teacher's thinking about what will be observed in a place value lesson, what is expected if the students are understanding the intended learning of the lesson, and what mistakes might be made.

Having collected and interpreted the information, the next step is to consider how the results will be reported to others. Information must be conveyed to the intended audience (students, parents, school administrators) in a way that communicates the intended outcomes of the assessment. This could be the students' mathematical knowledge, their dispositions towards mathematics, the ways in which mathematics is being used, and their ability to think mathematically. Readily understood formats for communicating this information are needed—descriptive narratives, profiles where grids indicate the goals of the assessment, grades, and so on.

Observe: blocks in right columns, using right language of place value, recording numerals in correct columns, correct naming of place value.

Mistakes to expect: recording 6 + 7 as 13 and leaving both digits in the ones column, not trading.

Place value lesson

Assessment tools

Even before considering what tools will be used, the first and most important consideration is to determine what is going to be assessed. If teachers want to know whether their students are able to communicate mathematical understandings or ideas, they need to select an assessment activity that will enable them to do this. Depending on the type of communication desired, this could require students to undertake an oral presentation, participate in a discussion during the lesson, prepare a written report or a journal reflection, or create an advertising brochure. These formats require different communication styles, and hence consideration of this is also necessary. If the requirement is to assess students' engagement, persistence and/or motivation in mathematics, the teacher might decide to implement a problem-solving task and, through the students' final products, as well as observations of students working and/or of journal reflections, be able to assess whether or not the desired characteristics are apparent.

The means by which students are expected to demonstrate their mathematical knowledges, attributes and dispositions must be conducive to facilitating the intended outcomes. It is unreasonable to assess when students have not been given ample opportunity to develop understandings of what is being assessed.

Once a decision has been made about what will be assessed and how best to engage students in activities that will facilitate what is to be assessed, then consideration needs to be given to how the data will be collected. For example, if the task is for students to demonstrate their understandings of subtraction with internal zeros, the teacher might ask them to send an email to a friend who is having trouble with the process. In this way, teachers can observe the students working through tasks. They can also listen to students' conversation with peers, or request the students to describe in their journals how they did the task. These activities will allow students considerable scope for demonstrating their understandings. The processes by which assessment will occur should be an integral part of the teaching episode—the lesson or unit.

● Observation

Perhaps the most common assessment tool in the primary school classroom is observation. Observations are non-intrusive forms of assessment whereby the teacher simply observes what students are doing and/or saying. Observation is used a great deal in the early years of schooling and becomes increasingly formal as students move through school. That is not to say that observation is not used in the upper primary grades; rather, it is not seen to be so popular in the years where there is an increasing reliance on traditional forms of assessment. However, in everyday interactions, observation is a key tool for judgement, and many teachers' 'intuitive' knowledge about students comes through this means. In the upper years, teacher observation can often result in a challenge to formal results on pencil-and-paper tests when teachers 'know' that the results are counter to their expectations.

To fully utilise observation as an effective assessment tool, teachers need to articulate what they want to observe and how they will record their observations. Being aware of what to observe is central to good observations. Identifying the key behaviours, language, and so on in the classroom is essential to the assessment process. The observations link with the intended outcomes of the teaching episode, the common errors that are likely to occur and other aspects that are a feature of the lesson (such as cooperation in groups, sharing, and so on). The teacher should have a good knowledge of these, as is evident in the planning process. The focus of the observations can vary from individuals through to groups, and can relate to any of the foci of the mathematics program.

One of the key features of good assessment through observation is keeping records of the observations. These typically come as anecdotal records. An effective method is the use of Post-it® Notes. These can be made for individual students and pasted into the individual's profile at the end of the lesson or day, enabling quick transfer of information. Alternatively, a flipboard can be made using file cards for each student. These are taped on to a clipboard, slightly overlapping, as shown in the diagram. Notes can be made on the cards, which when full are placed in the file card system. This makes for easy record-keeping. Journals can also be used and anecdotal notes taken as the teacher roams around the room. This system is easy to implement but transferring information from the journal to individual student records can be time-consuming.

Steffie is finally able to extend the AB pattern. She used Unifix counters for 4 patterns today. 3/5/02

Recording observations on Post-it® Notes

	Counting: May 13
John	*Basic OK, but difficulties with facts*
Mark	*Having difficulties with all skip counting. Only counts in ones*
Tim	*Has shown competence with all counting*
Jo	*Can do all counting tasks easily. Can do counting in 10s!*
Meg	*Has problems with 5s. Wants to revert to 2s*
Ny	*Problems with language*

Anecdotal observations

A further tool for observation is the use of running records that can be recorded using a checklist or similar class list record. The records can be checked off if using an outcomes list or specific list of expected behaviours, or through the use of anecdotal records against the student's name. For example, in the checklist illustrated in the diagram, the teacher has noted the expected outcomes for the teaching episode along the top of the checklist and the students' names down the side of the page.

Using a 'traffic light' system for recording observations, the teacher uses red to indicate no understanding, yellow to indicate some understanding and green to indicate full understanding. This simple system can be used to record observations in a quick and manageable fashion. In some cases, teachers even note the date of the observation. This is particularly useful when students' understandings appear to be developing.

Another tool for observation is more open-ended, whereby the teacher has a blank class list; as behaviours are observed, the teacher records them against the students' names. This type of observation schedule allows for more descriptive accounts to be entered and for the inclusion of observations that may not have been possible using the more prescriptive checklist with nominated outcomes.

All five formats allow for teachers to record their observations, albeit in very

Anecdotal observations may be made using a clipboard

	Count in 2s	Skip count	Count in 5s	Basic add'n facts
John	●	●	○	●
Mark	○	○	○	●
Tim	●	●	●	●
Jo	●	●	●	●
Meg	●	●	○	
Ny	●	●	●	○

Checklists with outcomes

different ways. The format selected can depend on the immediate need for the assessment (such as report cards), the personal style of the teacher and what the data will be used for in terms of teaching.

Performance tasks

As students are engaged in working on mathematical tasks, many of the skills that they are learning and developing can be assessed through observing them at work. For example, the ability to use a protractor to construct or measure an angle can be assessed through observation. In many cases, performance tasks resemble the tasks that would be undertaken in the world beyond school—measuring up a pool, building a frame for a window or tendering correct change.

● Consultation

Unlike observation, consultation involves some intrusion into the teaching/learning process. In most cases, the consultative process is simply posing questions to students and seeking their responses. It is often extended from observations, and can be used to complement the observation process. For example, while observing students working in small groups, an action may not be understood by the teacher so some questions are posed in order to gain more insights into the thinking of the students. Alternatively, when presenting instructions at the start of the lesson, quizzical looks on some students' faces may suggest they do not understand, so the teacher can interact with the students to clarify their concerns. Interactions with the students allow the teacher to gain a better knowledge of or insight into the students' thinking, problems or concerns.

● Focused analysis or questioning

Part of the consultative process is the posing of good questions, which takes time and effort. Sullivan and Lilburn (1997) argue that most of the questioning in mathematics classrooms falls within the lower regions of Bloom's taxonomy of questions (Bloom, 1956), and hence involves knowledge, comprehension and application. In contrast, few questions (even in major mathematics texts) lie within the regions of analysis, synthesis or evaluation. Sullivan and Lilburn (1997) define good questions as having three aspects:

1. They require more than remembering a fact or reproducing a skill.
2. Pupils can learn by answering the questions, and the teacher learns about each pupil from the attempts.
3. There may be several acceptable answers (p. 2).

In defining what makes a good question, Sullivan and Lilburn argue that good questions are useful for teachers when it comes to accessing and assessing what students know, since they create greater opportunities for students to display their understandings—or indeed their lack of understanding. Such questions move beyond mere memorising, whereby students demonstrate understanding of concepts rather than relying on recall of a technique. A common task, such as calculating the mean of a given series of numbers, often allows students to calculate the answer correctly as they know the procedure for such calculations without understanding the concept of 'mean'. Such questions are closed—that is, they have only one acceptable answer, and generally are restricted in the methods for resolution. In contrast, asking a question such as, 'Over the past six matches, a soccer team has averaged a score of 5. What could be the numbers of goals kicked in each of the matches?' allows for greater diversity in the responses possible and the logic or methods students will use to arrive at an answer. The opening up of questions in this way allows for greater flexibility by the students, as well as greater diversity in the responses and methods used, and allows teachers greater access to what students know and can do in a given topic.

Many teachers use questioning rather than direct instruction as the basis for their teaching. Being able to do this requires considerable skill, and enables the students to be engaged actively in the teaching episode while simultaneously providing the teacher with valuable information about what they are thinking. When planning teaching episodes, consider the form of the key questions to be used. The questions should move on from closed, low-level forms to include higher-order, open-ended questions. In considering the use of questions, some researchers in science (Lemke, 1990) and mathematics classrooms (Zevenbergen and Lerman, 2001) have found that particular questions—such as single response, yes/no questions—are used for controlling the content and flow of the lesson but have limited value in terms of intellectual engagement or substantive mathematical learning. Questions that encourage students to justify their thinking or working, propose different methods or solutions, or analyse their thinking, require a much deeper level of engagement, and hence more time for the formulation of responses; therefore, teachers need to practise patience when posing these questions to allow time for students to formulate responses.

Posing questions that demand more reflective or analytical thinking may involve less interaction across the class. To ensure participation from a wide range of students, a useful strategy can be to use the 'think–pair–share' strategy, whereby students formulate a response, share with a peer in pairs, then move to bigger groups—which could include groupings of four, eight and/or the whole class. This strategy is less threatening and allows students the time and space to work on their responses. Sharing with peers can also help to fine-tune, adjust or rethink their responses and thinking strategies.

● Open-ended questions

There are two main forms of questions—closed and open. A closed question provides some very focused questioning—such as, 'What is the mean of 5, 4 and 3?'—where there is only one answer, which students can get either correct or incorrect, and one identified method for solving the task. It yields little other information (other than using error analysis).

TEACHING IDEA

Open-ended questions

- Over a series of 6 days, Joe caught a mean of 8 fish. How many fish could he have caught each day?
- Three numbers add up to 31. What might they be?
- A rectangle has an area of 36 cm². What might its dimensions be?
- A solid has 6 faces. What shape might it be?

By contrast, Sullivan and Lilburn (1997) have provided numerous examples of open-ended questions in mathematics. They argue that open-ended questions typically do not have an obvious path of resolution, and have multiple answers. An open-ended question provides for greater diversity of responses across a range of levels of understanding. Consider the question, 'If 5 is the mean, what six numbers could make this mean?' Such a question allows students to provide a range of responses, which could include numbers that are all the same through to very large and small numbers, as well as common fractions and decimals. The range of responses offered by the students allows the teacher greater access to students' deep understanding of the concept of mean, as well as catering for the range of understandings across the classroom.

Examples of closed and open questions

Closed	Open
What is the area of a sandpit whose dimensions are 15 cm $\times$ 20 cm?	If a sandpit has an area of 300 cm², what might its dimensions be?
What is the sum of 7 and 8?	If the total of two numbers is 15, what might the two numbers be?

To change a closed question to an open one, a simple strategy is to look at the answer to a closed question—for example, 'the area is 300 cm²'—and then work backwards; given the answer is 300 cm², the students have to work out the dimensions. The goal of the activity (in this case, area) and the process (calculating using multiplication) are the key aspects of the task; these remain preserved and the task is inverted. There will be a number of correct and different answers to this question.

Open-ended questions provide a wide entry point, since the students can respond to the questions in ways that allow for their particular learning levels (Zevenbergen, 2001). Consider the following example where students were given the task: 'At 4.00 p.m. each day, the average number of people in each car on a busy suburban bridge is 2.5. At any one time, there will be 16 cars on the bridge. Draw what this might look like.' The drawings shown in the diagram illustrate the variety of responses offered by students in the upper primary grades.

The range of responses shows that some students have complex understandings of how to represent a mean of 2.5; in diagram C, a variety of cars is drawn with a diverse range of people in each car. Other students (diagram A) showed that they knew the total had to be 40 people, representing this as a simple pairing of cars with two and three people.

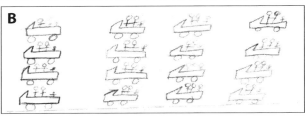

Student work examples of open-ended responses

Other students took the question literally and drew two people and half a person, or represented the half person as a small person (diagram B). These diagrams represent a range of different understandings of the concept of mean that would not be accessible through closed questions. All students have been able to participate and engage in the task, and to represent their understandings. These responses allow the teacher greater access to the ways in which students are thinking about the 2.5 as a mean. Unlike the conventional task, which would have been 'If there are 16 cars on the bridge and 40 people, what is the mean number of people per car?', the open-ended task allows students to demonstrate their understanding of the concept of mean, the ways in which they would calculate it, and how to represent it. It allows the teacher greater access to the knowledge that students have about mean which would not be accessible if the task were of the standard format. It allows for students to respond in ways that are appropriate for them, thereby allowing almost all students to respond, and demonstrate their current levels of understanding.

● Self-assessment and peer assessment

Most students have a good sense of what they know, what they have problems with and what they enjoy. Often they also know why this is the case. Yet teachers spend inordinate amounts of time and energy trying to access this information through a range of assessment instruments. However, there is little opportunity given to students to self-assess. Self-assessment allows students to disclose their mathematical understandings (or lack of). Developing a supportive ethos within the classroom is central to developing this aspect of assessment. This may take some time, since students have to learn to self-disclose; for most students, the practice of assessment in mathematics is to hide what you don't know rather than to show what you do and don't know.

Journal writing

Asking students to write in a personal journal about their mathematics learning, lessons and experiences can provide the teacher with valuable insights that otherwise would be inaccessible. Catalysts that help students write about the focus of assessment are an important consideration. Questions that can be posed include:

- What did I learn today/this week?
- What do I need more help with?
- What don't I understand?
- What did I enjoy about this week's mathematics?
- What would I like to learn more about in mathematics?
- What can the teacher do to help me more?
- What did I think the teacher did well (or poorly) in teaching mathematics?

These types of questions are useful as they allow students to discuss freely the issues they are confronting in the mathematics classroom.

A particular ethos needs to be developed in a classroom when undertaking journal writing. Depending on the ethos that develops, various forms of journal writing can be nurtured. For example, a teacher may decide that journals are a personal communication between the student and teacher only, so that quite personal information may be included. Alternatively, the journals may be a more open document so other information will be included (or excluded) depending on who will read the documents.

> I think I need more help with areas. I make mistakes when I have to convert units as I don't understand why the numbers get so big. I think that I should only divide by 100s when using centimetres but I don't and I don't understand why.

Example of journal writing in mathematics

Reflection

Most of the work in self-assessing requires students to be self-reflective and pose questions to and of themselves: 'Could I do this in a better way?' 'What have I learnt from doing this work?' 'Why am I doing this this way?' Reflecting on what and how they do things in mathematics, how they feel about mathematics, what they need to know more about and why can encourage students to recognise their strengths and weaknesses in mathematics.

● Work samples and portfolios

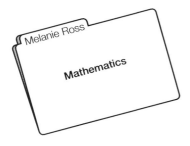

Example of student portfolio

The range of work undertaken by students in the mathematics classroom can be seen as representing their understandings of and/or dispositions towards mathematics. This work can be collected and collated into a systematic filing system so that it is seen to constitute pieces of evidence about the students' learning. The work samples should clearly show how they demonstrate learning. The pieces can include reports, diagrams, worksheets, problem-solving activities and the like. Some pieces may need to be annotated in order to clarify the reason for their inclusion.

A number of issues need to be considered in order to develop such a portfolio:

- Is the purpose of the portfolio to document mathematics learning?
- What will constitute evidence and who decides what will be included?
- Should the evidence represent full understanding, best work, developing ideas, drafts, etc?
- What is the best way to store and keep it?
- Will it be used for a particular purpose, or be ongoing?

While the traditional format for a portfolio consists of a manila (or similar) folder, with pieces of work and a page that indexes the contents, along with the point of the documentation, other formats can be used, such as a computer file (or disk), CD-ROM, book or video.

Student writing

One of the less commonly used formats for assessment is student writing. This can be in the form of a report or similar text. A catalyst can be used (such as those in the MCTP packages in Lovitt and Clarke (1992)) and the student writing used as evidence of learning. Consider the example in the Teaching Idea. Students are being asked to write a story about the number of jelly beans in the jar as represented by the graph. The task requires the students to interpret information from the graph and to communicate this to colleagues and peers. This demands a high level of communicative skills.

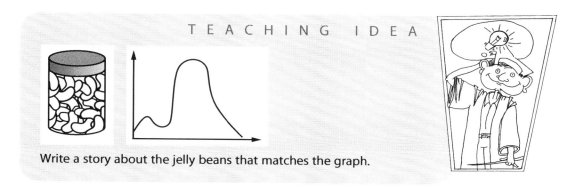

TEACHING IDEA

Write a story about the jelly beans that matches the graph.

Another strategy is to ask students to write a letter to a friend who was not at school and explain what they learnt in mathematics that day. This encourages reflective thinking and aids in the development of a language of explanation.

Tests

Tests do have a place in assessment, albeit a small one. Tests have limited value in terms of assessing what students know, feel and believe. Not only are tests difficult to construct well, they can be notoriously unreliable in testing what they are thought to be testing. A good assessment tool can be a student-constructed test. Such items serve two purposes: the construction of the test items by the students gives the teacher insights into what students know or can do; and the final test provides teachers with information about students' performance. They are useful for classrooms because, since all students have had some input into the construction of the items, the results should represent the diversity of understandings within a particular classroom. With this diversity represented in the test, all students should have success in answering questions. The tests are also useful in that they are owned by the students and will be less stressful since they themselves constructed the test.

Rubrics

0	Shows little or no understanding of place value. Can't trade. Subtraction incorrect.
1	Places numbers in right columns, trades but can only do in one column. Subtraction incorrect due to incorrect recording.
2	Place value recorded OK. Subtraction in tens but does not transfer to ones column.
3	Place value OK. Trading done correctly but records ones place value incorrectly when trading the 10 to 1.
4	Place value, trading and subtraction all completed correctly. Justifies response.

Specific marking rubric

A	Little or no understanding Cannot represent ½, ¼ or ⅕ numerically or pictorially.
B	Some understanding Can divide the unit into the number of nominated parts but does not understand the notion of equal parts.
C	Full understanding Can represent a range of fractions, can represent part-whole relationships, understands equal parts.

Simple rubric

One way in which you can document students' understandings is through the use of rubrics. These tools have a number of very good functions for teachers. By forcing the teacher to consider what will constitute evidence of learning, and the degree of learning, the rubric documents what the teacher sees as the critical elements of the teaching outcomes. The rubric should specifically target the key learnings for the teaching/learning process. This requires the teacher to be aware of the common errors that could be made, and what is expected if full understanding is evident.

The rubric in the diagram shows what the teacher is expecting to observe and how it will be recorded. The rubric rating scales are nominal scales. The scales are used to indicate the types of response given. Performance is scale-marked along a continuum so that the overall performance is rated. This rubric is quite specific, and is targeted at the intended learning outcomes of a very specific teaching episode.

Rubrics can be used in a more holistic way, be very detailed and contain up to ten rating scales, or be quite simple and involve only three rating scales. The scales generally are designed to describe the outcomes rather than to act as a scoring system. The detail provided in the rubric is designed to support the intended outcomes for the teaching episode, thus reinforcing the overall impetus for the lesson or unit. Teachers need to be very clear about the intended learning outcomes for the teaching episode in order to develop a marking rubric. At the simple level illustrated in the diagram, a rubric could be used for students' fraction knowledge to indicate little or no understanding, a developing understanding and full understanding. What constitutes these categories is the basis for the rubric, so the teacher needs to be fully cognisant of what is expected from the students.

Opening up assessment

The most common relationship in assessment practice is that it is undertaken by the teacher on the students for the purpose of reporting to parents, and usually through testing procedures. Clarke (1992) has strongly advocated that assessment networks be opened up to create better assessment practices, in pathways represented in the diagram below.

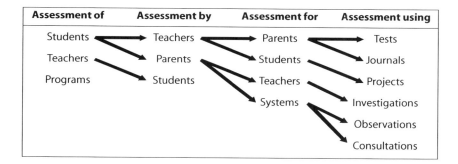

Assessment of	Assessment by	Assessment for	Assessment using
Students	Teachers	Parents	Tests
Teachers	Parents	Students	Journals
Programs	Students	Teachers	Projects
		Systems	Investigations
			Observations
			Consultations

● Monitoring and record-keeping

Keeping accurate records of student performance is a key task in assessment. Teachers need to develop techniques that will easily be incorporated into their work programs and styles of working. Some strategies have already been listed in the preceding sections, including a range of checklist formats and anecdotal records that can be noted in a general list or through a file card system. Keeping track of anecdotal records jotted while observing students at work can be difficult. A system that can work is the use of a folder where pages can be inserted (so that

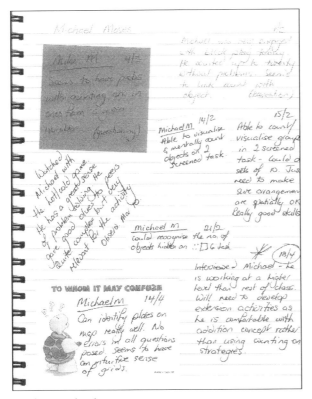

Teacher notebook

more pages can be added at any time). The teacher can use the document to scribble notes about the student, as a collection point for the Post-it® Notes noted in an earlier section, or for storing other information collected about the student. Having tabs at the side of the page allows the teacher to access any student very quickly. Such a document may not necessarily be neat, but it provides a common collection point for the various items that will be included.

Technology can also be a useful tool for record-keeping, particularly when using a quantitative program. Recording scores into this format can be simple and, once entered, scores can be easily calculated, plotted and made into reports for authorities and parents. There is also the handy option of inserting comments on the data entered into a spreadsheet (through right-clicking the mouse to come to the 'insert comment' command). This enables teachers to add supplementary notes on student performance.

		Test/16	Project/20	Test/20
		Feb-10	Mar-30	Apr-06
Jade	Scott	13	18	12
Melanie	Brown	11	17	18
Martin	Cumolov	12	19	18
Harley	Downer	7	11	19
Terri	Dunston	13	18	18
Paulette	Winston	12	16	17

Using a spreadsheet for record-keeping

Communicating results

Results of assessment need to be communicated to students, teachers, parents, administrators and increasingly to authorities. Test results tell any of these parties in a student's education what the student is capable of doing. They may be useful for monitoring large-scale systems, but have little other impact on learning.

Students can be informed about what they know, what they need to know and how they can improve their understandings. Qualitative information specifically targeted to the goals of the learning can support this. Communicating results to parents and caregivers, however, is often constrained by school policies and reporting formats. Most schools adopt some form of parent evening where individuals can discuss their child's progress with the teacher. This is often supplemented by a report card, usually at the end of each semester. Results are usually passed on to the teacher who will next work with the student—which could be within the same school setting or in another school. Part of the rationale behind the outcomes-based education that is dominating considerable reform in Western mathematics is the need for a standard language and format for monitoring and reporting student learning. Being able to state that a student is at a particular level in number, for example, and working with particular concepts within the number sub-strands, enables teachers to have a better understanding of what the student is able to do and understand. These results are often supplemented by qualitative and other data in a portfolio that moves with individual students as they progress through school. As the student moves through the mathematics curriculum, pieces of evidence are inserted and removed to show growth in learning and to provide evidence of how that growth has been demonstrated.

In terms of reporting to parents, developments within outcomes-based reporting include the use of software packages to support teachers in writing report cards. In some cases, it is simply a case of using a cut-and-paste mechanism whereby outcome statements and subsets of those outcomes—elaborations or indicators, as determined by the jargon of the initiative—can be taken from a database of the curriculum and inserted into the student's report card. This process documents achievement levels and moves away from the qualitative comments found on many report cards.

● Trends in reporting

Over the past few decades, changes have occurred in the ways that teachers report students' results. Sometimes teachers will be asked to present student information in particular ways due to parental (or employer) expectations. By understanding where interest groups are coming from and why they have particular expectations of teachers, teachers are able to justify their approaches to reporting. Changes in reporting over the past 30 years are summarised in the table.

Changes in reporting

Period	Type of reporting	Strengths	Weaknesses
Up to 1970s	Recorded scores, e.g. %₀ or 60%. Position in class	Gave rankings or position in class	Did not indicate what was known or not known
1970s to mid 1990s	Descriptive text, e.g. participates actively in discussions	Indicated strengths of the student (and sometimes weaknesses)	Did not indicate learning—what students did know
Current	Learning outcomes, e.g. operates on 2-digit numbers	Describes what students have learnt— in terms of outcomes stated in syllabus	Emphasis on learnt outcomes, not on attitudes or areas of growth

As this table indicates, there are pros and cons to most approaches to reporting. As with all teaching, mathematics teachers need to identify the reasons for reporting in particular ways. For example, while contemporary education focuses on reporting against learning outcomes, there are cases where test scores are featured. This is evident in the massive push by authorities to test students on statewide (or national) test schemes.

■ REVIEW QUESTIONS

6.1 Explain the relationship between teaching, assessing and learning, using an example to illustrate the points being made.

6.2 Outline and discuss three approaches to assessing learning other than pencil-and-paper testing.

6.3 What are the strengths and weaknesses of the approaches outlined in the previous question?

6.4 Develop and implement an open-ended question in a classroom. Bring examples of student work into class and discuss what you have learnt about students' learning.

6.5 What are some of the ways in which teachers can report student learning?

6.6 Visit the ACARA website and see if you can identify a relationship between ICSEA scores and numeracy scores.

Further reading

Clarke, D.J. (1992). Activating assessment alternatives in mathematics. *Arithmetic Teacher*, 39(6), 24–9.

Cole, K. (1999). Walking around: Getting more from informal assessment. *Mathematics Teaching in the Middle School*, 4(4), 224–7.

Ensign, J. (1998). Parents, portfolios, and personal mathematics. *Teaching Children Mathematics*, 4(6), 346–51.

Lee, C. (2001). Using assessment for effective learning. *Mathematics Teaching*, June, 40–3.

Parker, D.L. and Picard, A.J. (1997). Portraits of Susie: Matching curriculum, instruction, and assessment. *Teaching Children Mathematics*, 3(37), 376–82.

Schloemer, C.G. (1997). Some practical possibilities for alternative assessment. *Mathematics Teacher*, 90(1), 46–9.

Thompson, D., Thompson, K. and Else, N. (2000). Alternative assessment. *Australian Primary Mathematics Classroom*, 5(4), 29–32.

Tonack, D.A. (1996). A teacher's views on classroom assessment: What and how. *Mathematics Teaching in the Middle School*, 2(2), 70–3.

Wright, B. and Stewart, R. (1999). Can teachers know too much? *Australian Primary Mathematics Classroom*, 4(2), 4–7.

References

Australian Curriculum, Assessment and Reporting Authority (2010). NAPLAN. Available from <www.naplan.edu.au>.

Bloom, B.S. (1956). *Taxonomy of educational objectives: The classification of educational goals: Handbook 1, Cognitive domain*. New York: David McKay.

Clarke, D. (1992). *Assessment*. Melbourne: Curriculum Corporation.

Department of Education and Early Childhood Development Victoria (2006). Early Numeracy Research Project. Available from <www.education.vic.gov.au/studentlearning/teachingresources/maths/enrp/default.htm>.

Department of Education and Training Western Australia (2004). *First steps in mathematics.* Melbourne: Harcourt.

Ellerton, N.F. and Clements, M.A. (1997). Pencil-and-paper mathematics tests under the microscope. In K. Carr (ed.), *People, people, people: Proceedings of the 20th Annual Conference of the Mathematics Education Research Group of Australasia* (Vol. 2, pp. 155–62). Rotorua: MERGA.

Lemke, J.L. (1990). *Talking science: Language, learning and values.* Norwood: Ablex.

Lovitt, C. and Clarke, D. (1992) *The Mathematics Curriculum and Teaching Program.* Melbourne: Curriculum Corporation.

Pimm, D. (1991). Communicating mathematically. In B. Shire (ed.), *Language in mathematical education: Research and practice* (pp. 17–24). Philadelphia, PA: Open University Press.

Sullivan, P. and Lilburn, P. (1997). *Open-ended maths activities: Using 'good' questions to enhance learning.* Melbourne: Oxford University Press.

Wright, R.J., Martland, J. and Stafford, A.K. (2000). *Early numeracy: Assessing, teaching and intervention.* London: Paul Chapman.

Zevenbergen, R. (2000). 'Cracking the code' of mathematics: School success as a function of linguistic, social and cultural background. In J. Boaler (ed.), *Multiple perspectives on mathematics teaching and learning.* New York: JAI/Ablex.

Zevenbergen, R., Hyde, M. and Power, D. (2001). Language, arithmetic word problems and deaf students: Linguistic strategies used by deaf students to solve tasks. *Mathematics Education Research Journal,* 13(3), 204–18.

Zevenbergen, R. and Lerman, S. (2001). Communicative competence in school mathematics: On being able to do school mathematics. In M. Mitchelmore (ed.), *Numeracy and beyond: Proceedings of the 24th Annual Conference of the Mathematics Education Research Group of Australasia* (Vol. 2, pp. 571–8). Sydney: MERGA.

CHAPTER 7

Working mathematically

One of the most common descriptions of mathematics is that it is about doing pencil-and-paper calculations. This view is under challenge in current reforms in mathematics education. Contemporary mathematics education focuses on thinking and working mathematically so that mathematics becomes a way of seeing and acting in the world, rather than an activity in and of itself. To encourage students to think and work mathematically requires that they engage in a range of tasks, problems and investigations (Schoenfeld, 1992). The aim of this chapter is to present the notion of working mathematically in relation to problem-solving, connecting mathematics beyond the mathematics classroom and communicating mathematics through multiliteracies.

Working and thinking mathematically

In her study of mathematicians, Burton (2004) found that for those people working with mathematical concepts and processes, working mathematically is vastly different from the practices found in school mathematics. She argues very strongly that the practices used by mathematicians should be fostered in school mathematics so that students can learn how to work mathematically. Rather than rely on teacher direction, individual work and rote learning, Burton argues that a stronger focus must be placed on working collaboratively, talking about mathematical ideas. She maintains that arguing and talking about ideas, use of intuition and passion, and regularly going back to known knowledge to extend and build new knowledge are key processes that should be featured in contemporary classrooms.

Using many of the principles identified by Burton, Boaler (2008) similarly reports that quality teaching in mathematics classrooms included the use of group work so that students could talk, discuss, support and challenge ideas in ways that were far less intimidating than whole-class teaching. Boaler also suggests that teachers need to have tasks that create opportunities for high levels of mathematical understanding, that the teacher should take a role that enhances and supports learning (rather than directs learning) and that students need to have an opportunity at the end of the lesson to share their learnings with their peers. This final part of a lesson is one that encourages debate among peers but within a supportive environment. The development of this new model of teaching mathematics requires some time for students to learn the new protocols for teaching, but when they do the results are strong. Boaler's work shows that there are strong outcomes when the model of teaching moves from direct instruction to one that fosters deep mathematical learning by creating new spaces for working mathematically.

Working mathematically is an important life skill for effective living in the world beyond school. In order to be competent and effective citizens, students need to exit school with the dispositions and competencies in mathematics/numeracy that will allow them to participate fully in the activities they undertake. This means being able to use and work with mathematics in a way that empowers them in their everyday lives.

Contemporary approaches to mathematics education encourage students to think mathematically. Thinking mathematically refers to a disposition to use mathematics to solve problems and tasks in a manner that is logical and based on mathematical principles. Mathematics is a very useful tool to work through problems. Thinking mathematically is encouraged through the curriculum by posing problems that adopt and apply mathematical knowledge, skills and processes. The linking of realistic examples to mathematics is a key feature of teaching since students need to see the relevance and purposefulness of mathematics to real-world situations.

Problem-solving

The importance of problem-solving for fostering mathematical thinking gained prominence in the 1970s. It has been used in mathematics classrooms with greater or lesser effect ever since. For problem-solving to have most effect and benefit, it needs to be an integral part of the classroom ethos rather than simply an activity. More recently, problem-posing has been emphasised in addition to problem-solving (English, 1996).

A problem has been defined by Mason and Davis (1991) as something that gets inside the head of the learners so that they become motivated and challenged by the task or question. It is a question or task that does not have an obvious answer or a clear path for resolving it. Problem-solving is more than solving word problems such as 'Three birds were sitting on a fence, one flew away, how many are left?' For students who cannot determine that this is a subtraction task, this is a problem, but for most students it is little more than a routine task embedded in a word problem. Fostering mathematical thinking requires more motivating problems.

Teaching problem-solving requires teaching *through* problem-solving, and teaching *about* problem-solving. Teaching through problem-solving means immersing students in a variety of novel, challenging and motivating problems as a natural part of the mathematics program. Teaching about problem-solving scaffolds learning through the provision of strategies. However, students also need to have an adequate understanding of the mathematics involved in the problem before embarking on tasks (Pengally, 1989).

Polya's (1973) four-stage model for problem-solving is a generic strategy that underpins much of the teaching through problem-solving:

1. Understand the problem—what is the problem asking?
2. Devise a plan for solving it.
3. Carry out the plan.
4. Look back and reflect on the solution obtained.

In classrooms, this model is often referred to as 'see, plan, do, check'.

While the four steps are proposed as isolated units, Mason and Davis (1991) have suggested that this is not the case, and in fact the steps tend to blur. Students can move backwards and forwards between the steps as they attempt to work out ways to solve the problem. It is possible for students to become caught up in an endless cycle of reading the problem and thinking about ways to solve it, then rereading, to the point that they eventually give up without even attempting the problem. Students need to be supported to move beyond the blueprint of problem-solving to enable them to develop specific strategies for solving problems.

Problem-solving is assisted through having a wide repertoire of strategies from which to select when encountering different problems. Particular strategies suit particular problems. Knowledge of strategies is a product of experience and engagement in solving a variety of problems. The teacher's role is to provide such experiences. Common problem-solving strategies include:

- Create a table.
- Make a drawing.
- Think aloud.
- Act it out.
- Look for a pattern.
- Guess and check.
- Identify unwanted information.
- Use a simpler example.
- Identify other alternatives.
- Make generalisations.
- Work backwards.
- Check the answer.

In the following section these strategies are discussed in more detail.

● Create a table

- A carpenter constructs desks with four legs and stools with three legs. At the end of the day, she has used 43 legs. How many desks and stools have been made?

1 desk	2 desks	3 desks	4 desks	7 desks	10 desks
$43 - 4 = 39$	$43 - 8 = 35$	$43 - 12 = 31$	$43 - 16 = 27$	$43 - 28 = 15$	$43 - 40 = 3$
$39 \div 3 = 13$	$35 \div 3 = $ no	$31 \div 3 = $ no	$27 \div 3 = 9$	$15 \div 3 = 5$	$3 \div 3 = 1$
13 stools	no	no	9 stools	5 stools	1 stool

Using a table to organise information helps students to see patterns emerging and can help to identify any missing information. After the first four attempts, a pattern seems to emerge whereby every time the desk number increases by 3, there will be a remainder divisible by 3, which can result in the stool being constructed. Using a table allows the student to see a pattern beginning to emerge (every third time) and then to trial the idea (guessing and checking).

● Make a drawing

- If there are five people in a room and they all shake hands with each other, how many handshakes are there altogether?
- A frog in a well hops up 3 metres each day and slips back 2 metres each night. If the well is 10 metres deep, how long will it take for the frog to hop out?

Using drawings allows students to visually construct the problem. Some problems are best solved using this strategy. These include spatial problems (such as maps and paths). It allows students to 'see' relationships. When using this strategy, it is important to stress to students that they should not be spending time on details in the drawings but only providing sufficient information to demonstrate the problem. Doing this also supports students in learning what is the key information in a problem and ignoring redundant or irrelevant information.

● Think aloud

- As I was going to school, I met a teacher who had 24 students in her class. Each student had two siblings. Each sibling had two pets. How many teachers did I meet?

By encouraging students to think aloud, they are able to hear their verbalisations. This allows two processes—thinking and talking—to support students' problem-solving. Speaking their thinking processes supports communication as well as encouraging metacognitive processes.

● Act it out

- A party of 3 people goes to a restaurant. The dinner is $10 a head so it costs $30. As they are regulars, the manager decides they can have a $5 discount. The waiter decides it is too difficult to share $5 so gives each person a discount of $1 and keeps the other $2. The meals are now $9 which means that the dinner cost a total of $27. The waiter kept the other $2 which makes for a total of $29. Where did the other dollar go?

In this case, the students would be able to act out the process of the bill-sharing to see where the missing dollar went. By modelling the process, they are able to see the problem.

● Look for a pattern

- Square numbers are so called because of the pattern they form as the square 'grows'. How many squares are there when the square has a side of 5, or 10? How long will a side be when there are 81 squares in the square?
- A king has decided to reward his servant for his good deeds by offering a choice of whether he would like to take the wheat in each square multiplied by 10 or whether he would prefer to place a grain of wheat on a checkerboard and double the number in each subsequent square. Which option would you take and why?

Early mathematics involves considerable work on patterning, in number and spatially. This is a very important skill (and attitude) for competent thinking in mathematics, and hence should be continually encouraged and reinforced. Many mathematical discoveries have been made through people identifying patterns (such as Fibonacci's numbers). Often the pattern can be seen when the data are entered into a table; in other cases, such as spatial patterns, the patterning needs to be seen spatially.

● Guess and check

* Using the numbers 1 to 9, place them in the grid so that the numbers add up to 15 in any direction.

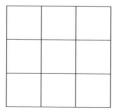

   The guess-and-check strategy is often frowned upon in mathematics, as if it were an inferior process. However, it can be useful and should be encouraged as one of many possible strategies. When using an educated guess, students need to rely on identifying the key information and some strategies for the resolution of the problem. Students should be encouraged to develop guessing strategies that are based on some knowledge and experiences rather than indulging in blind or wild guessing.

● Work backwards

* You spent $21 at the supermarket. The fruit cost twice as much as the meat, which cost twice as much as the chocolates. How much did the meat cost?
* A length of rope was cut in half to share between two landowners. One owner needed to use ⅔ of it for tying his dog to the shed. The piece that is left is 1 metre long; how long was the original rope?

   The working backwards strategy encourages the student to see what the answer is and then systematically work from that point to account for what has happened earlier.

● Identify unwanted information

* A new car purchase price is listed at $19 760. The fuel consumption is listed as 100 km on 10 litres. If the new owner drives an average of 1500 km per month, how many litres of petrol would have been consumed in a year?

Often mathematics tasks and problems are given to students so that they need to work with the numbers provided. However, in the world beyond school, such examples rarely exist so there is a need to recognise what is important and relevant information to solve a problem. Examples should be given to students that encourage this disposition to develop.

● Use a simpler example

• How many squares are there on a chessboard?

Often problems can be complicated by the size of the numbers or the nature of the patterns being used. By making the task smaller or simpler, reducing the size of the pattern or breaking the task into smaller components, the task can be seen as manageable. With the example of a chessboard, by reducing the original chessboard pattern into a smaller task—such as a 3×3 grid—the inherent components of the task remain unaltered and the student is able to manage the task. Through gaining insight into solving the smaller task, students can extend this learning to the larger or more complex task.

● Identify other alternatives

• The monkey needs to get to the bananas—how many ways can it get there? What is the shortest route to the bananas?

Bananas

Encouraging students to seek alternatives—either in the resolution of the answer or the process—allows them to generate new ways of seeing the problem and of how to solve it. A useful process for developing these outcomes is to promote group discussions—within a group and between groups. Allowing students to hear how others have solved the problem and the answers they have generated enables them to see alternative modes and responses to the same task. Once this ethos is a part of the classroom culture, it is useful for students to evaluate the responses their peers propose. Such a process should be undertaken with care so that a negative ethos does not develop—a constructive process should be nurtured. One way this can be achieved is through the SWOT process, where students identify Strengths, Weaknesses, Opportunities and Threats to the processes being used. Fostering this metacognition is an important aspect of working mathematically.

● Making generalisations

- When three consecutive numbers are added together their total is 3. What are the numbers?
- A rectangle has sides that are twice as long as its height. How long might the sides of a rectangle be?

When using this type of question, the way in which it is resolved can lead to a general rule for this type of problem. Focusing on the broader features of the problem (rather than the specifics) can result in students gaining deeper insights of greater significance than when they focus on the smaller elements of the specific task. Once students have gained proficiency with problems with numbers, they can start to generate general principles leading to algebraic thinking with problems that have no numbers.

● Check the answer

A most important strategy for students to learn is checking their answers. This enables them to identify errors—in the answer as well as the process. Using a strategy of reasonableness of the answer, students can be encouraged to see whether their answer is plausible. Estimating before working through a problem can be a useful process in that it allows students to see whether the answer they arrive at is reasonable.

Problem-posing

In contrast to problem-solving, problem-posing encourages students to create their own problems. Using the principles and strategies discussed above, problem-posing is seen to create significant opportunities for students to engage in mathematical ideas (English, 1997).

Moses et al. (1990) propose four principles to support students with problem-posing:

1. Focus students' attention on key information to be contained in the problem.
2. Commence with familiar concepts or ideas.
3. Encourage the use of ambiguity when designing problems as this opens up the problem.
4. Encourage students to set restrictions (or domains) to their problems.

The problem-posing approach is gaining a strong foothold in classrooms where teachers are attempting to make teaching more authentic.

Technology for thinking mathematically

The changes in the wider society towards a much more technology-orientated lifestyle have resulted in significant changes in the ways mathematics is developed. There is a substantive body of knowledge that demonstrates how computers and calculators encourage much deeper levels of mathematical thinking than is possible with pencil-and-paper work. For example, in her work with calculators, Groves (1995) has shown that students can gain a greater sense of number in the early years through guided play with calculators. Whereas most curriculum documents demarcate the first year as being related to number study of 1 to 10 or maybe 1 to 20, the following year with numbers to 100, and the subsequent year with numbers to 1000, Groves has shown that students in their first year of school can gain a strong grasp of numbers up to 1000. In part this is due to the reduction of other demands—such as writing numerals—and a greater emphasis on thinking mathematically. Similar studies have shown that graphic calculators have enormous benefits in terms of gaining deep understandings of functions and algebra (Doerr and Zangor, 2000). This is due to the capacity of the calculators to show the graphs and how the gradients and intercepts change with changes to the function. This dynamic process allows students to see the effects of changes quickly; old methods focused on physically plotting the graphs, which meant that few could be undertaken, thereby restricting the dynamic visualisation process.

The computer and its applications have had similar effects on mathematical thinking, fostering deeper understandings of mathematical ideas. For example, the graphing function allows students to painlessly construct bar graphs or pie graphs. Constructing pie graphs by hand is very tedious. The spreadsheet function allows students to explore which graphs are best for what purposes in ways not possible (or at least restricted) when constructing them through pencil-and-paper methods. Similarly, the spreadsheet function requires students to develop simple instructions for calculations—such as add cell B1 to cell B2. This process requires some algebraic thinking and provides a strong rationale for using algebra.

It becomes possible to pose deeper questions about mathematics when using technology—questions about patterns and place value, as well as higher order questions such as 'Why', 'What if?', 'What happens when?', 'Why is this happening?' The focus shifts from the process of construction to understanding the reasons, purposes and rationales for undertaking particular work.

Young people in modern society have grown up in a technology-rich society. Their familiarity with technology as a tool makes it amenable to considerable innovations in mathematics education. Software such as spreadsheets can be open-ended, with considerable potential for using it in many ways to support various aspects of mathematics. Other packages tend to be closed and restricted to particular topic areas. This includes specific packages such as the software packages for learning multiplication facts or other operations (Maths Blaster®). Such packages can simply be upmarket worksheets, and hence encourage rote learning rather than mathematical thinking. Packages such as LOGO® or Geometers Sketchpad® may be orientated towards particular topic areas, but still have an open-endedness to them, allowing students to explore other aspects of mathematics—particularly working mathematically. When selecting software, teachers should be cognisant of what they contain and critical of their shortcomings.

Making connections

For too many students, the experience of school mathematics is not positive—they perceive it to be a difficult and irrelevant subject. The task of the teacher is to create meaningful and purposeful connections between mathematics and other spheres of life and school. That is, mathematics needs to be transdisciplinary—it must link realistically to other curriculum areas—for students to see why they study mathematics. That it informs many other areas is a connection rarely made explicit to students. It is not possible to study social studies without a knowledge of number, area, graphing, location or statistics. It is not possible to study science without a knowledge of number, exponentials, measurement, or recording and interpreting data.

Mathematics applies to all areas of life. Without a deep understanding of mathematics, people's lives are impoverished, and in many cases severely restricted. Not being able to undertake calculations can affect salaries, budgets and banking; not being able to measure accurately can affect such activities as cooking, gardening, handyman work, and so on. Mathematics is an integral part of quality life, yet considerable numbers of school-leavers exit with very negative experiences and self-concepts of themselves as users of mathematics.

Teachers need to make connections between mathematics and other curriculum areas and activities beyond the school. These activities should not be tokenistic but instead need to be realistic. In many early years classrooms, a common activity is that of the class shop, where students practise shopping through purchasing and operating with money. Students see that these are not real activities but pseudo-activities, where the shopping activity is a mask to justify the mathematics. Making connections so that students can see the value of mathematics demands that realistic activities are undertaken.

The value of activities such as purchasing the items on a shopping list becomes meaningful, relevant and purposeful when the items are realistically priced and there is some culminating activity—such as going shopping to purchase goods, or making a class party where foods are purchased and/or cooked. When the activity resembles the activity undertaken in the world beyond school, there is every opportunity for the students to begin to make the connections between school mathematics and the wider world. They can see why they need to study mathematics and that mathematics is not an irrelevant area of study.

Making connections happens on three levels:

1. *Within mathematics*—where links are made between the various strands of the curriculum. It is not possible for number study to exist in isolation from other areas. For example, area relies on a knowledge of multiplication or repeated addition. These links need to be developed.
2. *With other discipline areas*—where connections are made with other curriculum areas—such as science, health and physical education, and social science.
3. *In worlds beyond schools*, where the mathematics is placed in realistic contexts. These contexts demonstrate how mathematics can be used to support and enhance problem-solving.

Communicating mathematically

Current thinking in the area of literacy involves the four different roles of the reader (Freebody, 1992). This model of multiliteracies provides a strong link to the roles of the numerate student. More than simply considering mathematics teaching as linking language to concrete and symbolic representations, multiliteracies encompasses mathematical reading, representing, recording and communicating mathematically. The role of the reader is not passive, but integral to the construction and interpretation of mathematical 'texts', including concrete materials, photographs, pictures or books, video or non-static representations, teaching episodes where there is considerable interpretation of teacher actions by the students, and the traditional written formats of textbooks, worksheets and board work.

Students talking about representations

Consider this photograph. A number of interpretations can be made of what is happening here. Is one boy helping the other? Is he telling the other he is wrong? Right? Many interpretations are possible.

In terms of multiliteracies, the mathematics classroom is a text about which students will make interpretations (or readings). When teaching is seen in this way, it becomes possible to understand the learner as a much more active participant in the classroom, and in so doing allows the teacher to realise that students can construct very different interpretations of what has been said or done. This moves the emphasis away from seeing students as giving right or wrong answers to one where the role of the teacher is to understand why students construct responses and understandings in the ways they do. Not only are the communications related to mathematics, but so also are the texts within which the mathematics is being conveyed to the students. Meaning-making becomes multidimensional.

By applying the multiliteracies perspective to mathematics, communicating mathematically becomes more than simply recording with pen and paper. Communicating mathematically encourages several different types of communication:

- *Oral communication*—where students 'talk' mathematically. This can be in open discussions or small groups. The teacher sets tasks that require students to talk about the mathematics they are using, the ways in which they are working, the answers they are constructing. This form of communication can include plays and other performances.
- *Visual communication*—this can be in the form of displays which may be flat, two-dimensional representations such as written projects, or three-dimensional displays such as working projects (e.g. constructions).
- *Digital communication*—creating displays using computer technology. This could include the use of specific programs such as students talking about representations spreadsheets to show patterning (or pre-algebraic thinking); or PowerPoint presentations to create a visual display that documents learning.
- *Textual communication*—the traditional format where students write mathematically. This can be useful for explanations, justifications, refutations, conjectures, and so on. This form of communication is particularly important in developing students' metalanguage of mathematics and encouraging them to think mathematically. Students can be asked to write their own mathematics questions, or to reflect in journals about what they have learnt or had difficulty

with, or to write a letter to a friend explaining what they have learnt in mathematics. A commonly used example here has students writing a story about a graph depicting the height of water in a bath. The students' stories indicate their understanding of graphs, and their interpretation of data, slope, gradient, and so on.

• *Symbolic communication*—the unique feature of mathematics is its symbolic form. To communicate in this truncated language, students need to know the specific language of symbols. The use of symbols allows mathematics to develop a particular form of communication. For example, students are introduced early to the signs $+$, $-$, $\times$, $\div$ and $=$; these are followed in later years by $<$, $>$, $\leq$ and $\geq$, among others. While these symbols are an integral feature of mathematics, they also are a key to a specific language. Common usage of these signs can result in misconceptions being developed. For example, early addition experiences focus on a language such as 3 apples and 4 apples makes 7 apples, so that the equals sign comes to be seen as equivalent to 'makes'—4 and 3 is equivalent to 7. The language becomes inappropriate in the case of 1 and 1 makes 2, since 1 and 1 can make 11 when the two numbers are placed side by side or 7 if placed at right angles!

Jenny has $62 and micheal has $9. Jenny lost $19 or how much do they now have together

If I had $26 & my sister had $46 how much money is there all together

I had 60¢ and my friend gave me $2.50 how much do I have Now?

Miss Kelly went to the shop and brought 12 apples for 3.00 each How much did she spend? _____

Examples of students' own mathematics questions

■ REVIEW QUESTIONS

7.1 Why is working mathematically a strong emphasis in current practice in mathematics teaching and learning?

7.2 Identify six strategies for problem-solving. Briefly outline each strategy and discuss when this strategy is most useful.

7.3 Identify three modes of communicating mathematically. Provide an example of each mode and how it would be used in the classroom.

7.4 Why is thinking mathematically an important skill to develop in students?

Further reading

English, L.D. (1997). Promoting a problem-posing classroom. *Teaching Children Mathematics*, 4(3), 172–9.

Jitendra, A. (2003). Teaching students math problem solving through graphic representations. *Teaching Exceptional Children*, 34(4), 34–9.

Lovitt, C. and Clarke, D. (1990). *Mathematics Curriculum Teachers Package 1*. Melbourne: Curriculum Corporation.

——(1990). *Mathematics Curriculum Teachers Package 2*. Melbourne: The Curriculum Corporation.

References

Boaler, J. (2008). Promoting 'relational equity' and high mathematics achievement through an innovative mixed-ability approach. *British Educational Research Journal*, 34(2), 167–94.

Burton, L. (2004) *Mathematicians as Enquirers*. Dordrecht: Springer.

Doerr, H.M. and Zangor, R. (2000). Creating meaning for and with the graphing calculator. *Educational Studies in Mathematics*, 41(2), 143–63.

English, L. (1996). Children's problem-posing and problem-solving preferences. In J. Mulligan and M. Mitchelmore (eds), *Children's number learning* (pp. 227–42). Adelaide: Australian Association of Mathematics Teachers.

English, L.D. (1997). Promoting a problem-posing classroom. *Teaching Children Mathematics*, 4(3), 172–9.

Freebody, P. (1992). A socio-cultural approach: Resourcing the four roles of the literacy learner. In A.J. Watson and A.M. Badenthorp (eds), *Preventing reading failure*. Sydney: Ashton Scholastic.

Groves, S. (1995). The impact of calculator use on young children's development of number concepts. In R. Hinting, G.E. FitzSimons, P.C. Clarkson and A.J. Bishops (eds), *Regional collaboration in mathematics education* (pp. 301–10). Melbourne: Monash University.

Mason, J. and Davis, J. (1991). *Fostering and sustaining mathematics thinking through problem solving*. Geelong: Deakin University Press.

Moses, B., Bjork, E. and Goldenberg, E.P. (1990). Beyond problem solving: Problem posing. In T.J. Cooney (ed.), *Teaching and learning mathematics in the 1990s (1990 Yearbook)* (pp. 82–91). Reston, VA: NCTM.

Pengally, H. (1989). Becoming mathematical problem solvers. In B. Doig (ed.), *Everyone counts* (pp. 1–5). Melbourne: Mathematics Association of Victoria.

Polya, G. (1973) *How to solve it.* Princeton, NJ: Princeton University Press.

Schoenfeld, A.H. (1992). Learning to think mathematically: Problem solving, metacognition, and sense making in mathematics. In D.A. Grouws (ed.), *Handbook of research on mathematic teaching and learning* (pp. 334–70). New York: Macmillan.

CHAPTER 8

Early number

Early number study has tended to take a sequential approach. Typically, in the first year of schooling the focus was on the numbers 1–10, followed by exploration of the teens. Numbers 20–99 were studied through a place value approach. The simultaneous linking of linguistic, concrete and symbolic representation of the numbers studied underpinned all work in this approach. This sequence has been questioned because it may not capitalise upon students' intuitive counting and number understandings. Research has continually highlighted the importance of developing number sense in the early years and building upon the intuitive knowledge that students bring to school (Anghileri, 2001a; Beishuizen and Anghileri, 1998; Clarke et al., 2006; Perry et al., 2008). This chapter presents early number from a number sense perspective using a contemporary approach to the teaching of this foundational topic.

Number sense

One of the aims of school mathematics is to develop students' number sense. Number sense relates to a level of comfortableness and familiarity with numbers—an 'at-homeness' with numbers. It is about understanding number meanings, knowing relationships between numbers, knowing the size of numbers, and knowing the effects of operating on numbers (Anghileri, 2006; McIntosh et al., 1992). Number sense is never complete; developing number sense is a lifelong process that is promoted through many and varied experiences with using and applying numbers.

Before they start school, many young students exhibit sophisticated levels of number sense that often go unnoticed by teachers. Take the example of a pre-school child attending a school-entry interview. The mathematics task presented to the child was to take a packet of twelve coloured pencils and make two groups. After much deliberation, the child separated the pencils into a group of seven and a group of five. The child was then prompted to make two equal groups. After further deliberation and much consternation, the child removed two pencils from the larger group and held them in her hand under the table. This was clearly not the expected response, and the child displayed a level of uncomfortableness with her action. Upon questioning, her reasons for the indecision became apparent—the first separation of pencils was according to 'boy' colours (black, dark brown, dark blue, dark green, grey) and 'girl' colours (yellow, orange, pink, red, purple, light green, light blue). The second separation was the removal of light green and light blue from the 'girl' colours, but because they were girl colours they could not be included in the boy colours group. The child's response shows the level of intuitive knowledge of division and the depth of cognition that this task invoked.

Becoming aware of the mathematics knowledge and number sense young children bring with them when they commence formal schooling is vital to ensure that students maintain an interest in and eagerness towards school mathematics. Researchers' investigations into early number development (e.g. Clarke et al., 2006; Fuson, 1988; Gelman and Gallistel, 1978; Steffe et al., 1983) have provided research-based evidence of the depth of young children's intuitive mathematics knowledge—that is, knowledge acquired without formal instruction.

Once in the school situation, the teacher's task becomes one of capitalising on the number knowledge that students bring to the learning situation. Teaching strategies would include focusing on counting activities; using calculators to explore big numbers and patterns within numbers; working flexibly with numbers; verbalising their thinking and solution strategies; visualising and articulating patterns in numbers; recognising group size without counting; developing number benchmarks (fives and tens); learning basic facts; relating size of numbers to a meaningful context; and creating an awareness of the effects of operating on numbers. The priority for early number work is to build upon and promote number sense.

Pre-number

Many young children display an awareness of number: they know how old they are; they know how many siblings they have; they know their parents' ages; they know what television channel their favourite show is on; they know the floor their mother's office is on; some even know their telephone number. Many children can count beyond 10 before they commence school. To promote understanding of number and the development of counting skills is to capitalise on children's early number knowledge and to provide further, explicit activities relating to the processes of classifying, patterning and subitising.

● Classification

Being able to classify objects into groups is a prerequisite for number study. In order to be able to count objects in a group, students first need to identify the objects themselves. Consider the collection of shapes in the diagram.

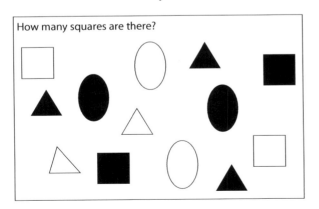

How many squares are there?

Asking students to count a particular item in a collection requires that they can identify the object to be counted. If they are unable to distinguish the circles from the squares, then counting the squares is not possible. Classification activities can arise from many collections found in the classroom, the home, the playground or the environment. Buttons can be sorted according to colour, size, shape, texture, number of holes; toys can be sorted according to purpose, type, number of hands, colour of skin, and so on. Encourage creative thinking about classification of a collection. Classification skills relate not only to mathematics, but also to other subjects, including science, art, music, social and environmental education.

● Patterning

Patterning is a key aspect of mathematics. By providing students with quality experiences in patterning, they will be more likely to see patterns in problem situations, and be able to make generalisations about their patterns, which is the basis of algebraic thinking.

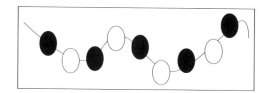

Patterning activities include the following:

- *Copying patterns*—students are given a pattern and their task is to copy it. This can be a string of beads or an array of blocks, but the task for the students is to simply copy what is displayed.
- *Identifying the next item*—from a given pattern, students are asked to predict the next item(s).
- *Extending patterns*—students are asked to extend the full pattern.
- *Creating patterns*—this is creative pattern work where students construct their own patterns.
- *Determining the missing element in a pattern*—a pattern is given, but one of the elements in the pattern is omitted. The student's task is to identify what is missing.

Patterns do not need to be limited to beads, blocks and counters. They can be created through using the students themselves (one child standing, one child sitting, one child standing . . .). Patterns are also found in fabric (curtains, sheets, shirts, bedcovers). Encourage students to see patterns around them, to experience patterns in a variety of forms, before they are asked to create their own.

Subitising: Group recognition

Before they begin formal schooling, many young students have the capacity to subitise. Subitising is being able to identify the number in a collection or group without counting. This skill is very accurate for collections that number between one and five items. The skill of subitising is essential for the development of concepts of number (von Glasersfeld, 1982).

Subitising is a useful skill for early number work as it saves time in counting and it is often more accurate; it promotes number sense of the size of numbers (2 is less than 3 is less than 4) and assists simple mental addition and subtraction. It is important as it de-emphasises a counting-by-ones strategy for determining the number of objects in a collection. Instead, emphasis is on the automatic recognition of the number of items in the collection so as to free up cognitive space. This expedites subsequent work with operations and promotes flexible thinking about numbers and number combinations. Teachers of early years classes need to determine the extent of their young students' subitising skills, and build upon or promote their development.

Flashing small collections of counters on an overhead projector using flash cards, or using small collections on boards, is useful for this purpose. The arrangement of collections for subitising must be considered in such activities.

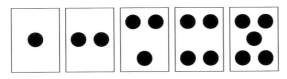

Many students rely on counting-by-ones strategies for addition and subtraction. Subitising reduces the need to rely on such strategies. For example, when students are asked 'How many?' when presented with a group of three and a second group of four objects, through subitising they automatically recognise the collections of 3 and 4—without counting—then mentally combine the groups to give the total.

Counting

Contemporary approaches to early number recognise and emphasise the intuitive counting knowledge students bring to school. Students are often familiar with numbers beyond 10. Old approaches often saw the teen numbers as problematic, but contemporary research suggests that, through experience, young students intuitively know teen numbers—often because of family members' birthdays and similar situations. Through playing with the language of numbers, students develop an aural sense of number names. Experimenting with number pattern names, students often count in tens as 'sixty, seventy, eighty, ninety, tenty, eleventy'.

Difficulties often emerge with the study of numbers when formal study of place value is introduced. Recent approaches to teaching number shift the emphasis away from place value to the intuitive counting strategies students bring to school, and develop as they progress through school.

Many young children can 'count' to 10 before they start school, but what is really happening is that they are merely repeating a sequence of words they have learnt, just as if this sequence were a poem or a nursery rhyme. Counting is not just being able to say number names in order. To distinguish between, on the one hand, merely saying the number word sequence and, on the other hand, using counting in a problem-based situation, Steffe et al. (1983) use the term 'counting' to refer to situations in which each number word is coordinated with an item. The items might be seen or might be visualised (i.e. imagined) (see also Steffe, 1992; Wright, 1991).

An important goal for the teacher is to determine the sophistication of a student's counting strategy beyond oral recitation of number names.

● Counting stages

In the literature, there are two main approaches to describing children's development of counting strategies. The approach developed by Steffe et al. (1988) has been applied extensively to classroom teaching and also in intervention programs (Wright et al., 2000, 2002), and is summarised in the table.

In contrast, many curriculum documents identify a less in-depth approach, and nominate three distinct stages of counting that can be observed among young students. Each stage represents progressively more refined counting skills.

Rote counting

This is when students can say a string of number names, and this string is very stable. The string may include all numbers 1 to 10 in the correct

Model for stages of early arithmetic learning

Stage 0: Emergent counting	Cannot count visible items.The child either does not know the number word sequence or cannot coordinate the number words with items.
Stage 1: Perceptual counting	Can count perceived items but not items which are screened (i.e. concealed).This might involve seeing, hearing or feeling items.
Stage 2: Figurative counting	Can count the items in a screen collection but counting typically includes what adults might regard as reductant activity. For example, when presented with two screen collections, told how many in each collection, and asked how many counters in all, the child will count from 'one' instead of counting on.
Stage 3: Initial number sequence	Child used counting on rather than counting from one, to solve addition or missing addend tasks (i.e. $6 + x = 9$). The child may use a count-down-from strategy to solve removed items tasks (i.e. $17 - 3$ as 16, 15, 14—the answer is 14) but not count-down-to strategies to solve missing subtrahend tasks (e.g. 17–14 as 16, 15, 14—the answer is 3).
Stage 4: Intermediate number sequence	The child counts-down-to, to solve missing subtrahend tasks (e.g. $17 - 14$ as 16, 15, 14—the answer is 3). The child can choose the more efficient of the count-down-from and count-down-to strategies.
Stage 5: Facile number sequence	The child uses a range of what are referred to as non-count-by-ones strategies. These strategies involve procedures other than counting by ones, but may also involve some counting by ones. Thus, in additive and subtractive situations, the child uses strategies such as compensation, using a known result, adding to 10, commutativity, subtraction as the inverse of addition, aware of the '10' in a teen number.

Source: Wright et al, 2000, p. 26

sequence, or some numbers in the sequence may be omitted (e.g. 1, 2, 3, 5, 7, 9, 10). Students at this stage appear to be counting but are not able to come to the correct amount of 'how many' objects are in a collection.

Point counting

At this stage, when given a collection of objects to count, students will make gestures of counting by touching objects in the collection, but cannot tell how many. For a collection of five objects, students may count to 8 as they assign two numbers ('1, 2') when they point to the first object, and two more numbers ('3, 4') to the second object. Characteristically at this stage, when asked again to state how many, students begin again at 1, stating there are '1, 2, 3, 4, 5, 6, 7, 8' objects.

Rational counting

Students are deemed to be rational counters when they demonstrate understanding of the following counting principles, as identified by Fuson (1988):

- Each item that is to be counted is assigned one (and only one) number name.
- The order in which items are counted must be undertaken in a specific and appropriate number order. That is, items must be counted in the order of 1, 2, 3, 4, 5, 6 . . .
- The order in which items are counted is irrelevant—it is possible to start at either end of a collection.
- The last number counted represents the number of items in a collection.

Young students counting collections of items

● One-to-one correspondence

The term one-to-one correspondence is often used to signal students' progress in counting. One-to-one correspondence is evident when students can match a count with an object. One-to-one correspondence relates to the first principle of counting listed above.

Conservation of number

The influence of Piaget on mathematics education was described in Chapter 2. Piaget's work on number provided us with the term *conservation* in relation to mathematics. From his research, Piaget suggested that the ability to see that the number of objects in a collection remains the same, regardless of how it is arranged and rearranged, depends upon the level of cognitive development of the child.

In the diagram, students may think that the bottom row holds more blocks than the top row. The inability to see that the number remains constant in spite of its spatial arrangement indicates that students are focusing on attributes other than 'how many' objects are in the collection. When students can recognise that the number of objects remains constant regardless of how they are organised, they

are said to be able to *conserve* number. The implication of this is not that teachers should delay all counting activities until students can conserve number, but rather that they should provide students with many and varied counting activities where collections are rearranged in a variety of patterns and arrangements and encourage students to verbalise their thinking and mental strategies. Investigating the extent to which students can conserve number provides valuable information for building an early number program full of variety and with plenty of rich experiences and language.

● Counting strategies

When students are rational counters, they need to be immersed in activities that promote the development of more effective and efficient strategies for counting. Three main strategies are counting on, counting back and skip counting. The effective use of the counting strategies is a precursor to the development of the four operations (addition, subtraction, multiplication and division), depending on the counting strategy. Students need to be immersed in a range of activities in which counting strategies can be applied so that they develop flexible thinking when working with numbers and also utilise number sense.

Counting on

Counting on is where students begin with a given number and count on from this. In order to use this skill, students need to be able to 'break' the counting number string at any point by being able to determine what number comes after, or what number comes before, any given number. Being able to break the string means that the child does not have to start counting at '1' to determine the next number forward (or backwards) from any given number. This is an important skill to apply to addition and subtraction problems.

Two counting-on strategies are useful for simple addition:

1. Counting-up-from is where two numbers are given, the student identifies the larger amount and counts-up the smaller amount starting from the larger number, e.g. 3 + 8. 'Begin with 8, and then count 9, 10, 11. So the result is 11.'

2. Counting-up-to is where the starting point and the end-point are given and the task is to find out the difference by keeping track of how many counts are made. This often involves finger counting, and this should be encouraged. For example, 8 and what is 11? 'Begin with 8 and then count to 11, so that is 9 . . . 10 . . . 11. I counted three numbers so the answer is 3.' In this type of problem, one of the numbers (or one part of the amount) is given, as well as the total amount. The task is to find out the other number (or the other part of the amount) to equal the total. As the amounts (the two parts) are given and they are added, they are referred to as addends. In this particular problem, only one addend is given, so it is referred to as a missing addend problem.

Counting on

Counting back

Counting back

Counting back is the reverse of counting on. Students often experience more difficulty with this aspect of counting than with counting on. It is of similar difficulty for young students as it is for adults when they attempt to say the alphabet backwards. Counting back requires being able to identify 'what comes before' any given number. It is important for the development of the subtraction process.

Two counting-back strategies are useful for simple subtraction:

1. Counting-back-from is where two numbers are given and the student counts back the smaller amount beginning with the larger number, e.g. 11 take away 3. 'Begin with 11, and then count back 10, 9, 8. So the result is 8.'
2. Counting-back-to is where the starting point and the end-point are given and the task is to find out the difference by keeping track of how many counts are made. This often involves finger counting, and this should be encouraged. For example, I started with 11 and there

are now 8, so how many were removed? 'Begin with 11 and then count back to 8, so that is 10 . . . 9 . . . 8. I counted three numbers so the answer is 3.' As with the counting-on problems where the unknown quantity was within the problem, this type of subtractive task is known as a missing subtrahend problem.

Skip counting

Skip counting is counting in multiples of any other number apart from 1. Skip counting is promoted as students listen to the patterns and observe the rhythm of counting in various multiples. For example, counting in tens is a simple pattern as each count after 10 ends in 'ty'. For fives, the 5 and 'ty' endings are heard in a repetitive chant. After students have skip counted orally, record the numbers in the skip for students to explore. They will see how the end-digits in the number start to take on a repeated pattern. For example, when counting in fives, they will note that each number ends in either 5 or 0. When they skip count in fives starting at, for example, 2, they will see that the repeating pattern is 2, 7, 2, 7. This helps number sense and understanding of divisibility rules. Skip counting forwards and backwards represents an important learning experience for students as it is the precursor to multiplication and division. For example, to find out what six groups of 5 totals, students can skip count in fives—5, 10, 15, 20, 25, 30. Similarly, to find out how many groups of 2 there are in 10, students would begin at 2 and skip count to 10, keeping track of how many counts they make.

● Opportunities for practising counting

Students should be given plenty of opportunities to practise counting. Where possible, links to the world beyond school should be encouraged. Ask students whether they watch the counter at the petrol bowser when they get petrol with their parents and what patterns they notice. Ask them whether they watch kilometre markers on the side of roads when they are travelling on long trips. Ask them whether they take the stairs two at a time and whether they count in twos while doing it. Ask them to think of devices that count in the real world. What patterns do they notice on a calendar?

Students should also be given opportunities to practise counting (particularly skip counting) using technology such as calculators. Many calculators have an inbuilt constant key so that when 2 + 6 is pressed followed by the = key, the answer will appear. Pushing the = sign will keep adding 6 to the last figure. This allows students to see skip counting in action and to experience larger numbers than would be possible if they were to undertake the skip counting themselves.

Number benchmarks

The capacity to be flexible with numbers means knowing how numbers work, how they are related to other numbers, and how they can be rearranged and regrouped. Such knowledge supports mental computation and is a major component of number sense. Being able to relate numbers to other numbers is about building number benchmarks. Common number benchmarks are 5 and 10 (and later, multiples and powers of 10). When numbers become benchmarks, they provide a reference point to use for calculation or conceptualisation purposes. For example, to calculate 9 + 4, a common strategy is to think of 9 as 10, and then add 3. Because the relationship of 9 being one less than 10, the strategy of adding 1 to 10 and reducing the 4 to 3 is applying flexible thinking around 10. Knowing the combinations that make 10 (1 + 9, 2 + 8, 3 + 7, etc.) is also using 10 as a benchmark, particularly when adding two-digit numbers. For example, when adding 38 and 74, the first strategy might be to recognise that 3 and 7 is a combination that makes 10, even though in this example these two numbers are 30 and 70 and the total is 100. Five is a benchmark by seeing it as half of 10 (and later 50 as half of 100), but supporting children to see the number combinations that make 5 (1 + 4, 2 + 3) helps promote 5 as a benchmark in its own right that can be used as a reference point in other situations.

To build familiarity with 5 and 10 as benchmarks, and hence promote number sense, students need be to provided with a range of materials and activities to experience the size of numbers and make connections between numbers. Teaching strategies need to be enacted that will enable students to develop these benchmarks to facilitate more effective counting and operation strategies. Encouraging children to count on their fingers, and hence develop finger patterns, takes account of the strategies and knowledge students bring to school (Fuson, 1988).

Subitising (the capacity to recognise the number of objects within a collection without counting) can be facilitated through the use of counters where a collection of counters is 'flashed' before children and they need to state how many are in the collection. Different arrangements of counters totalling 5 will build 5 as a benchmark. Educators recognise the value of counters to support the development of early number knowledge (Steffe et al., 1983). The reforms from the Netherlands also encourage the use of the abacus and strings of beads (Gravemeijer, 1994).

Five frame

3 and 2 more make 5
3 is less than 5

Ten frame

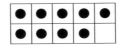

9 is one less than 10
9 is close to 10
9 is almost double 5

Reinforcing our base 10 numeration system can occur through promoting 5 and 10 as benchmarks, and this is supported through the use of five frames and ten frames. A five frame presents five counters in a line, with one counter placed in each cell of the five frame. Students use the five frame to show the combinations to 5 using two different coloured counters (5 + 0, 4 + 1, 3 + 2, 2 + 3, 1 + 4, 0 + 5). A ten frame is a double five frame and is used in a similar way to show combinations to 10. The ten frame has the added feature of being able to show that 10 is two groups of 5, or five groups of 2 (if displayed vertically). It also clearly shows that 5 is half of 10. The frames allow students to develop spatial patterns for numbers.

● More and less

When exploring numbers and relationships between numbers, the words 'more' and 'less' may often be used (Is 6 more or less than 7?). Increasingly, research indicates that the use of these terms in early number work can create difficulties for some students because of their unfamiliarity with the term 'less'. Rarely are children asked whether they want 'less' of anything; rather, they are asked, 'Do you want more ice-cream?' or 'Do you want more toast?' Teachers need to spend a little extra time when undertaking early number work that has a language component to it in order that all students have equal access to the intended learning.

Types of numbers

There are three different types of numbers that students will encounter in mathematics: cardinal, ordinal and nominal. It is not necessary for students to learn these names, but they need to be exposed to the three different types in their study of number.

- *Cardinal numbers* refer to the amount. Cardinality refers to finding 'how many'. Through early counting activities, students are immersed in cardinal numbers.
- *Ordinal numbers* refer to the consideration of 'where or which one'. Ordinality refers to a position, or the order of an object in a series—for example, the first car in the race, the third house in the street or the fifth rung on a ladder. Often the position relates to cardinal number and can relate to equal quantities separating items (such as rungs on a ladder) or unequal quantities between positions (such as a race). This can be an issue when students are using number lines and/or rulers and the misuse of rulers/number lines shows this misconception. For example, some students will use their ruler to add numbers (e.g. 5 and 7). Due to poor number sense, the strategy used will be a counting-on strategy, which will be misapplied. The student will locate the number 5 on the ruler/number line, and count on from that point, starting with the number 5. By the time the student counts on seven numbers, he/she is at number 11. The student is counting the line markers in the ruler and not taking into account that each distance between the line marker is the actual value of the number. The student is misapplying a rote counting strategy to a concrete representation that they do not fully understand. Students often experience difficulty with this aspect of number study, so experiences need to be organised to facilitate appropriate understandings to develop. Often students come to school with early ordinal numbers (first, second and third). A further difficulty with ordinal language is that many of the words used for ordinal numbers are also used for fractional concepts (e.g. fifth, sixth) so the distinction needs to be taught explicitly to students in order to avoid confusion.
- *Nominal numbers* refer to number names. Some objects—such as racing cars, postcodes, telephone numbers—have numbers as their

names. These types of numbers are useful for identifying important information but do not have the same properties as cardinal and ordinal numbers. For example, a baseball player may wear a shirt with the number 18 on it, but this does not mean that she is 18 years of age, or that she is the 18th person in the team. This number is for identification purposes.

Students need to have a range of experiences with these types of numbers. Early work with number typically is through cardinal numbers so students operate on such numbers. Students need to realise that they cannot undertake operations with ordinal or nominal numbers. For example, it is not possible to add two postcodes, or two places in a race, to arrive at an answer.

● Learning about numerals

Through a range of experiences with counting, students come to see the symbolic form of a number. This is referred to as the numeral. As they count beyond 9, students learn the patterns for combining numerals to make representations of larger numbers. Through relating the number name for the items counted and the symbolic representation, students intuitively learn about place value. There is some debate as to how place value is best taught—whether as a formal process or learnt intuitively. What is perhaps more important is that teachers know of different approaches so that when one approach is not successful they have other ideas of how to approach the concept.

● Writing numerals

As students come to develop a sense of number and cardinality, they also need guidance for writing numerals. There are 10 numerals that make up our numeration system. Depending on location, there are some particular ways of writing the numerals. In most countries with an Anglo-Saxon background, a similar format tends to be used (see diagram). However, with increasing globalisation, recognition of European numerals also needs to be acknowledged. For example, in most European countries, the numeral 7 is written with a bar through the upright, and 1 is written with a lead-in 'tail'.

1 2 3 4 5 6 7 8 9 0

Explicit instruction must be given to ensure students learn to write the numerals correctly, as displayed above, and as symbol formation is a part of handwriting, the class handwriting time should include a focus on correct numeral formation. As a result, students will value the importance of correct writing of numerals and see this as a key component of good handwriting. This will ensure that mathematics time focuses on conceptual development rather than symbol development.

Variety in exercises for promoting good symbol development is important to ensure students' interest and concentration are maintained. Suggested experiences include tracing over numerals already written on paper, drawing on a partner's back and then asking the partner to identify the numeral, placing the numeral on an overhead projector and projecting its image on a wall for students to trace, cutting numerals out of sandpaper for students to trace with their fingers, having students trace numerals on an overhead transparency with a water-soluble pen, making class numerals as in the photo, and making big numerals on the floor using rope or other thick cord. Students should be encouraged to talk through the writing process. For example, to make a seven, 'Start at the top, then you go across the top and then you go on a diagonal across to the bottom on the other side.' As students gain confidence in counting and writing numerals, they need to have experiences of counting on from the numbers given and writing them, as well as inserting numerals into a sequence where numerals are missing. This is often more difficult than counting on, so support needs to be offered to students when they undertake this aspect of writing their work.

Students creating numerals

Numeration and place value

● Numeration

Developing numeration is developing students' understanding of the Hindu-Arabic number system, which is the system of numbers in our society. The Hindu-Arabic system is based on five principles. A student would be expected to appreciate all five principles if understanding numeration:

- *Base*—the system is a base 10 system, and uses only ten digits (0, 1, 2, 3, 4, 5, 6, 7, 8, 9). These ten digits enable us to create any number.
- *Position*—the position of a digit in a number dictates its value, and every place in the system is related to other places by powers of 10. To understand the meaning of a number, students use the position of the digit. For example, in the number 435, the symbol 4 means that there are four hundreds as it is in the hundreds place, or position. Similarly, the 3 is in the tens position and the 5 is in the ones position.
- *Multiplicative*—the value of each digit is multiplied by the value of its place (in 435 the value of the 4 is 4 $\times$ 100, the value of the 3 is 3 $\times$ 10, and value of the 6 is 6 $\times$ 1).
- *Additive*—digits in a number are added to give the total amount (435 is 400 + 30 + 5).
- *Odometer*—the Hindu-Arabic system is a counting system in that every place counts. Thinking of how an odometer on a vehicle works helps with this notion. On each reel on an odometer are the digits 0 to 9. On each count, the reels turn—the 1s reel turns at every count; the 10s reel on every tenth count; the 100s reel on every hundredth count, and so on. This principle assists in counting and mental computation.

● Developing place value

The Hindu-Arabic system is a place value system. The development of place value understanding begins through exploration and recording of two-digit numbers. To promote sense of numbers beyond 10, students need to be immersed in activities to experience the size of numbers and to develop the principles of numeration listed above.

Bundling icy-pole sticks into groups of ten with a rubber band and holding ten bundles in the hand is an excellent way to feel the size of 100. When students say, 'I'd never be allowed to eat this many icy-poles', they are demonstrating a strong sense of number in relation to 100. Bundles of ten sticks also provide a physical referent when counting in tens. For numbers greater than 100, ten bundles of ten sticks can be fastened with a rubber band to further promote a sense of the size of numbers. The icy-pole sticks are a 'grouping' material, as they can be grouped and ungrouped with the use of a rubber band. Students can use this material to explore numbers beyond 10, demonstrating how they consist of tens and ones. For example, 35 is made up of three groups of ten sticks and five loose sticks. By unbundling one of the groups of ten, students can see that 35 is also made up of 20 and 15. Unbundling another group of ten shows that 35 is the same as 30 and 5.

In the photograph, a range of materials has been used to represent 35. All materials are proportional since the base 10 longs are equivalent to ten ones; a stick of Unifix blocks is the same as ten single blocks; a bundle of sticks is the same as ten single sticks. Representing place value in these forms encourages students to see the relationship between the representation and the materials.

Another common material for developing numeration and place value is through base 10 blocks. These blocks are also referred to as MAB or Dienes blocks (Zoltan Dienes created the blocks in the 1960s). The letters MAB stand for multibase arithmetic blocks, as they were created in a series of bases and not just base 10. However, MAB other than in base 10 are not a common sight in schools, and usually when teachers refer to MAB, they are referring to the base 10 MAB blocks. A set of base 10 MAB consists of ones, tens, hundreds and thousands blocks that can be used to represent numbers. For example, 453 would be represented with four hundreds blocks,

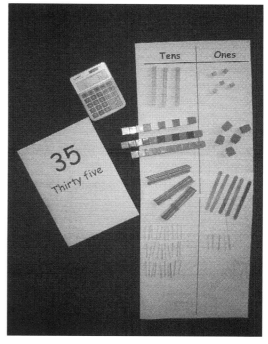

Ways to represent 35 with various materials

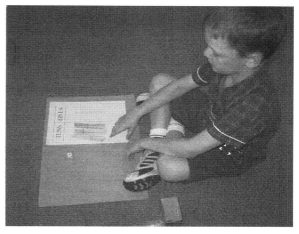

Using MAB to represent two-digit numbers

five tens blocks and three ones blocks. MAB cannot be pulled apart or reassembled; rather, they need to be traded. MAB are a 'trading' material as groups of ten blocks of one type are traded for another block.

● Symbolising numbers

When numbers are represented with concrete materials, the placement of objects may not reflect the way the digits are recorded. For example, the number 42 can be represented with base 10 blocks by arranging four tens and two ones blocks on the desk in any order. To relate the value of the blocks to the placement of digits in symbolic form, a place value chart is useful to organise the material. Relate the digits in the symbolic recording to the concrete representation.

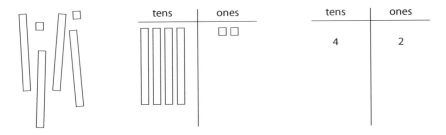

● Grouping and regrouping

To emphasise relationships between numbers and build place value understanding, students need a lot of experience with grouping and regrouping numbers. Using bundling sticks and place value charts, students can explore how numbers can be grouped and regrouped, and thus develop flexibility in thinking about numbers. With concrete materials, the number 34, for example, can be seen as consisting of three tens and four ones; with regrouping, it can also be seen as two tens and fourteen ones; and even as 34 ones. Flexibility with numbers lays the foundation for mental computation, and for understanding the written procedures for addition and subtraction.

Being able to regroup numbers into various component parts is evidence of good number sense. The process of swapping or trading ten items in a collection for one of the next item (e.g. MAB ten ones for one ten) to develop place value knowledge has been used for many decades with differing levels of success. Since the mid-1980s, researchers (e.g. Cobb, 1991; Treffers, 1991) have questioned the effectiveness of using base 10 blocks to teach place value.

● The empty number line

An alternative approach to developing children's place value knowledge is the Realistic Mathematics Project (RME) from the Netherlands. In this approach, the focus is on using a number line, not expressly for developing place value, but rather for developing mental computation around addition and subtraction, and using children's counting and number knowledge to support working flexibly with numbers. Through this approach, students' place value knowledge has been seen to develop without explicit teaching (Gravemeijer, 1994).

Two main strategies for working with the empty number line are the jump and the split strategy. The jump strategy is where a two-digit addition problem is given and the empty number line is used to track the situation. For example, in 28 + 39, the 28 is written on the empty number line and 'jumping' in 10s occurs three times to arrive at 58 (which means that 30 has been added to 28). To add the remaining 9, there is a jump from 58 to 60, followed by a jump of 7, to arrive at the solution of 67. A split strategy is where the number is broken into numbers that fit with the other number so as to add to 10. Using a split strategy for 28 + 39 means that the 39 is split into 2 + 30 + 7, and this is the way it is represented on the number line. The split strategy encourages students to work flexibly

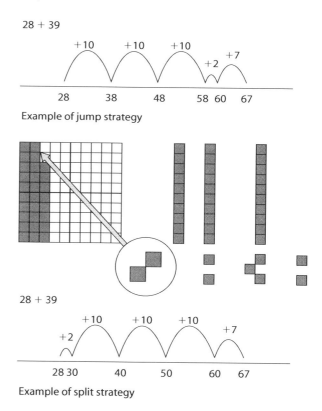

Example of jump strategy

Example of split strategy

with numbers by splitting them in ways that make them more usable, depending on the context of use. Combinations of jump strategies and split strategies can be used.

Another strategy is that of compensation, where the number may be rounded up and then the rounding up (or down factor) adjusted. For example, in the case here, the student may recognise that 39 is 40 – 1, so add (using jump, split or combination strategies) to the 28 (28 + 40) and then take the one away so that it is 68 – 1. Through these strategies, students come to see the effectiveness of the tens and ones, they use 10 as a number benchmark and they come to experience numeration informally so that the introduction of formal place value study builds on their intuitive understandings. The Netherlands reforms have drawn heavily on the intuitive methods of mental computation to show that, in the world beyond school, this is the predominant mode of thinking when adding (or working with) numbers.

● Base 10

An important aspect of understanding place value is knowing the relationship between each place in the Hindu-Arabic system. Each place is related to the place next to it by a power of 10. Explorations with a calculator can assist comprehension. Enter a number on a calculator, then progressively multiply this number by 10. Watch as the digits move progressively across each place, with zeros appearing to fill the gaps where the digits previously were located:

$$24 \times 10 = 240$$
$$24 \times 100 = 2400$$
$$24 \times 1000 = 24\,000$$
$$24 \times 10\,000 = 240\,000$$

This knowledge is much more powerful than the blindly applied strategy of 'adding zeros'. By progressively dividing by 10, the number becomes less, and the digits move back to their original position (or even become decimal numbers). Base 10 understanding assists in mental computation and understanding of the process of multiplication and division (see Chapter 10 on written algorithms.)

● The Number Framework

A Number Framework (Ministry of Education, New Zealand, 2008) was developed as a result of extensive research and professional development in 2002–07. The Number Framework provides a summary of knowledge and strategies required by children to operate successfully in the number strand. The framework emphasises the importance and inter-connectedness of knowledge and strategies. New number knowledge is a result of using advanced strategies, but strong knowledge is essential for broadening a student's strategy repertoire. The framework presents nine global stages, further categorised as relating to the counting stage or the part-whole stage.

The counting stages are:

- emergent
- one-to-one counting
- counting from one on materials
- counting from one by imaging, and
- advanced counting.

The part-whole stages are:

- early additive part-whole
- advanced additive part-whole
- advanced multiplicative part-whole, and
- advanced proportional part-whole.

The global stages sub-categorised as counting and part-whole reflect children's ever-increasing knowledge and understanding of number. The part-whole stages are signified by a student's capacity to 'recognise numbers as abstract units that can be treated simultaneously as wholes or can be partitioned and recombined' (p. 4). Number sense and thinking flexibly about numbers, as outlined in the preceding sections of this chapter, align children's developmental movement between the counting and part-whole stages in the New Zealand Framework. The part-whole stages reflect children's increasing knowledge and strategy development around number, moving from whole numbers to rationals. Teaching to promote the development of rational number knowledge and understanding is detailed in Chapter 11. It should be noted, however, that while the counting stages of the Number Framework have been validated by research, the part-whole stages have not (Young-Loveridge and Wright, 2002). Nevertheless, the stages provide a valuable way of thinking about children's number knowledge development and the growth of strategic thinking.

● Big numbers

As students' understanding of place value develops, conceptualisation of the size of numbers being dealt with must be promoted. As each new place in the Hindu-Arabic system is encountered, students need to investigate:

- How big is 1 000?
- How big is 10 000?
- How big is 100 000?
- How big is 1 000 000?

The Teaching Idea lists some questions that can be posed to students to encourage them to think about big numbers. An excellent way to visualise 1 million is to create a cubic metre from metre rulers. Place four 1 metre rulers on the floor and then ask for student volunteers to hold four more 1 metre rulers vertically at each corner of the square metre on the floor. Then use four more 1 metre rulers to finish the skeletal construction of the cubic metre. Place one MAB one cube (1 cm³) in the cubic metre and ask students to state how many cubic centimetres can be contained within the cubic metre. Gradually, students will realise that, with dimensions of 100 cm × 100 cm × 100 cm, the cubic metre holds one million cubic centimetres.

TEACHING IDEA

Big numbers

Estimate:

- How many hairs you have on your head.
- The number of words in a 200-page book.
- The number of people who would fit on the deck of the *Titanic*.
- What you could buy for $1 million.
- How many days/hours/minutes you have been alive.

- The number of grains in a cup of sand.
- How much water you would use in a shower over 1 year.
- How many centimetres of fencing would be needed for a paddock.
- How many kilometres a rider would ride when training for a bike race such as the Tour de France.
- How many grains of rice are in a kilogram bag.

Reading big numbers

The Hindu-Arabic system enables counting to continue beyond any given number. Reading big numbers is assisted by knowing that each three places (beginning at the ones) is a period, and that each period has its own label.

billions	millions	thousands	ones	thousandths	millionths	billionths
TEN BILLION BILLION	HUNDRED MILLION / TEN MILLION / MILLION	HUNDRED THOUSAND / TEN THOUSAND / THOUSAND	HUNDRED / TEN / ONE	tenth / hundredth / thousandth	tenthousandth / hundredthousandth / millionth	tenmillionth / hundredmillionth / billionth

The periods are signalled by a comma or a space. The periods of numbers are indicated in the table. The period names assist in naming very large numbers and also decimal numbers. Reading numbers occurs by considering the digits within each period as hundreds and then referring to the period in which that group of digits occurs. The following are two examples of how large numbers are read:

- 459 346 278: four hundred and fifty-nine *million*, three hundred and forty six *thousand*, two hundred and seventy-eight
- 6 589 320 400: six *billion*, five hundred and eighty-nine *million*, three hundred and twenty *thousand*, four hundred.

Once students see this structure, they enjoy the challenge of reading very big numbers.

Addition and subtraction

Through number and counting activities, students develop skills in simple addition and subtraction. Yet understanding of addition and subtraction is more than the skill of being able to add two or more groups together, or being able to take away an amount from a larger amount. Informal counting activities and the development of more complex counting strategies are precursors for operating on numbers. A rich conceptual understanding of addition and subtraction is necessary in order for students to be successful problem-solvers in situations involving addition and subtraction.

● The language of addition and subtraction

For many students, addition is narrowly conceptualised as 'joining', and subtraction as 'take away'. When presented with problems, words such as add, join, meet, altogether, arrive, take away, subtract, remove, eat, run away and leave will signal whether the required operation for solution is addition or subtraction. The strategy of looking for cue words (joining words and/or take-away words) is precarious, as it will not assist in all cases. Consider the following word problems:

- I went to the bank and withdrew $50. This left $100 in my account. What was my original balance?
- I added 50 grams of currants to my cake mixture, which made a total of 750 grams of dried fruit. How much dried fruit was in my mixture before I added the currants?

In these examples, the words 'withdraw' and 'add' are used, which are subtraction and addition words respectively. Yet the operations required to solve each problem are addition and subtraction respectively.

Students need to experience a variety of addition and subtraction problems that do not follow a standard format. A lot of research has been undertaken in this field, resulting in classification of addition and subtraction problems under such labels as compare problems, change problems, combine problems, part-part-whole problems and unknown start problems (see Chapter 3). While the labels are possibly confusing, the salient point of such research is that, in the real world, addition and subtraction problems are not packaged in simple 'join' or 'take away' language.

● Types of word problem

A simplified classification system that works well for both addition and subtraction problems involves labelling problems as joining, comparison, inaction or take-away problems. The labels provide some indication of how the problem is modelled with concrete material. Joining problems have a joining action, inaction problems are where no action takes place, comparison problems are where there are two groups to be compared, and take-away problems have a take-away action. The table provides examples of these problems for addition and subtraction.

	Addition	Subtraction
Joining	I have 5 mini cars. My mother gave me 4 more. How many do I have now?	I had 7 mini cars and I gave some to my brother. Now I have 3. How many did I give to my brother?
Inaction	I have 6 soccer cards in my pocket and Lisa has 3. How many soccer cards do we have between us?	I have 9 soccer cards in my collection. I have my favourite 3 in my pocket. How many are at home?
Comparison	I have 6 lollipops and Ben has 2 more than me. How many lollipops does Ben have?	I have 4 lollipops; Sam has 2 less than me. How many does Sam have?
Take-away	I ate 6 cherries, now I have 7 left. How many did I have to start with?	I had 14 cherries and I ate 7. How many do I have left?

Examples of four types of addition and subtraction problem

When working with addition and subtraction word problems, concrete materials may assist with representation of the elements within the problem, or acting out the situation may assist in meaning. The important thing to remember is that the problem-solving process where students are engaged in discussion, justification, verification and/or reasoning is the most valuable aspect. Immersion in a variety of problems with different language structures promotes familiarity with addition and subtraction problems and leads to greater mathematical understanding (see Chapter 3 for further details about word problems, language, equity and mathematical access).

Multiplication and division

Just as exposure to a range of addition and subtraction word problems promotes conceptual understanding and problem-solving skills, this is also the case for multiplication and division.

● Multiplication problems

There are four basic types of multiplication problem: grouping, rate, scalar and cross-product problems. Some examples are as presented below.

- *Grouping:* There are 3 bags that have 5 sweets in each bag. How many sweets altogether?
- *Rate:* Sara bought four lollipops that cost 5 cents each. How much did it cost her?
- *Scalar:* Jan has three times as many baseball cards as Peta. Peta has eight cards. How many cards does Jan have?
- *Cross-product:* Brian has three shirts and two pairs of shorts. How many different outfits can be created?

It is important that children be exposed to the range of multiplication problems so that they can experience the language of multiplication situations. Concrete materials support understanding of the problem and awareness of the multiplicative aspect of different types of problems.

For the grouping problem above, students put five counters (or blocks) in three specific piles. They use skip counting in fives, or recall the basic fact, to attain the solution. For rate problems, students need to put a coin next to each of the four lollipops. They then count up the value of the coins, or recall the basic fact, to reach the solution. For the scalar problem, two groups of material are required, each designated clearly as to their owner. This is actually a ratio situation where the two quantities are compared multiplicatively. For every card that Peta has, Jan has three cards. These cards should be displayed so the child can see that Jan has 24 cards when Peta has eight. It is important that the arrangement of the material matches the problem.

Cross-product problems are the most difficult multiplication situations. The best way to support students' understanding of cross-product solutions is to act out the problem using real or simulated materials. For the problem above, provide students with cut-out pictures of red and black shorts, and white, green and yellow shirts. Students then arrange the shorts and shirts together on the table to ensure that they have accounted for all possible combinations. They will come to realise that there are $2 \times 3 = 6$ possible combinations, and that in general you multiply the number of choices to find the solution. Organising this information in an array can assist in also representing and conceptualising the cross-product situation. The information in the array can be organised either way as shorts $\times$ shirts or shirts $\times$ shorts, as seen in the diagram.

Shorts	Shirts		
	White	Green	Yellow
Red	R, W	R, G	R, Y
Black	B, W	B, G	B, Y

Shirts	Shorts	
	Red	Black
White	W, R	W, B
Green	G, R	G, B
Yellow	Y, R	Y, B

● Division problems

There are two types of division problem, partition and quotition, classified according to the action that occurs when acting out, or modelling, the problem to assist with interpretation. Some examples follow:

- *Partition:* There are twelve lollipops and three children. How many lollipops will they each get?

- *Quotition:* There are twelve lollipops and each child is allowed three. How many children can the lollipops be shared between?

The difference between these two types of problems is the action of sharing. In partition division, the number of groups is known but how many will be in each group is unknown, so a dealing or one-at-a-time sharing action is invoked. In quotition division, the amount in each group is known, but the number of groups is unknown, so that amount is taken each time to determine how many groups will be made of the specified amount.

● Meaning and concrete representations

In multiplication, the basic meaning is grouping, and making groups can assist in representing multiplication situations. As students gain experience with a variety of multiplication problem types, organising the group into arrays also provides a structure for showing the commutative nature of multiplication (e.g. that 8×5 is the same as 5×8) and, depending upon students' facility with basic multiplication facts, can assist in determining the solution. For division, the basic meaning is sharing. Eventually, as students' understanding of multiplication and division grows, they will dispense with materials for representing situations and come to rely on their knowledge of basic facts or computational methods to solve the problems they encounter.

■ REVIEW QUESTIONS

8.1 Describe why 5 and 10 are benchmarks. What other number benchmarks might be useful for supporting number understanding? Give three examples of when 5 or 10 could be used as a benchmark.

8.2 Write a definition of number sense. Compile a list of questions that could determine whether (a) a young child, and (b) an adult has good number sense.

8.3 Compile a sequence of activities for skip counting that starts with students exploring skip-counting patterns in large numbers. Justify your approach.

> **8.4** What is subitising? What activities promote this skill? How does awareness of subitising inform your approach to teaching addition?
>
> **8.5** Compile a list of activities, including websites, that promote conceptualisation of large numbers (10 000, 100 000, 1 000 000).

Further reading

Auriemma, S. (1999). How huge is a hundred? *Teaching Children Mathematics*, 6(3), 154–9.

Baroody, A. and Benson, A. (2001). Early number instruction. *Teaching Children Mathematics*, 8(3), 154–8.

Bobis, J. F. (2002). Is school ready for my child? *Australian Primary Mathematics Classroom*, 7(4), 4–8.

Clements, D. (1999). Subitizing. What is it? Why teach it? *Teaching Children Mathematics*, 5(7), 400–5.

Cotter, J. (2000).Using language and visualization to teach place value. *Teaching Children Mathematics*, 7(2), 108–14.

Gervasoni, A. (1999). Teaching children the conventions of our number system. *Prime Number*, 14(2), 11–14.

Martin, B. (2000). What is counting? How do we teach it? *Prime Number*, 15(1), 5–7.

Taylor-Cox, J. (2001). How many marbles in the jar? Estimation in the early grades. *Teaching Children Mathematics*, 8(4), 208–14.

Wright, R.J., Martland, J. and Stafford, A. (2000). *Early numeracy: Assessment for teaching and intervention.* London: Paul Chapman.

Wright, R.J., Martland, J., Stafford, A. and Stanger, G. (2002). *Teaching number: Advancing children's skills and strategies.* London: Paul Chapman.

References

Anghileri, J. (2001a). Intuitive approaches, mental strategies and standard algorithms. In J. Anghileri (ed.), *Principles and practices in arithmetic teaching: Innovative approaches in the primary classroom* (pp. 79–94). Buckingham: Open University Press.

——(2006). *Teaching number sense* (2nd ed.). London: Continuum.

Beishuizen, M. and Anghileri, J. (1998). Which mental strategies in the early number curriculum? A comparison of British ideas and Dutch ideas. *Educational Research Journal*, 24(5), 519–38.

Clarke, B., Clarke, D. & Cheeseman, J. (2006). The mathematical knowledge and understanding young children bring to school. *Mathematics Education Research Journal*, 18(1), 78–103.

Cobb, P. (1991). Reconstructing elementary school mathematics. *Focus on Learning Problems on Mathematics*, 13(2), 3–32.

Fuson, K. (1988). *Children's counting and concepts of number.* New York: Springer-Verlag.

Gelman, R. and Gallistel, G.P. (1978). *The child's understanding of number.* Cambridge MA: Harvard University Press.

Gravemeijer, K. (1994). *Developing realistic mathematics education.* Utrecht, The Netherlands: CD-B Press.

McIntosh, A., Reys, B.J. and Reys, R.E. (1992). A proposed framework for examining number sense. *For the Learning of Mathematics*, 12(3), 2–8.

Ministry of Education, New Zealand. (2008). *The Number Framework.* Wellington: Ministry of Education.

Perry, B., Young-Loveridge, J., Dockett, S. and Doig, B. (2008). The development of young children's mathematical understanding. In H. Forgasz, A. Barkatsas, A. Bishop, B. Clarke, S. Keast, W. Seah and P. Sullivan (eds), *Mathematics Education Research Group of Australasia: Research in mathematics education in Australia 2004-2007* (pp. 17–40). Rotterdam: Sense Publishers.

Steffe, L.P. (1992). Learning stages in the construction of the number sequence. In J. Bideaud, C. Meljac and J.P. Fischer (eds), *Pathways to number* (pp. 82–98). Hillsdale, NJ: Lawrence Erlbaum.

Steffe, L.P. and Cobb, P., with von Glasersfeld, E. (1988). *Construction of arithmetic meanings and strategies.* New York: Springer-Verlag.

Steffe, L.P., von Glasersfeld, E., Richards, J. and Cobb, P. (1983). *Children's counting types: Philosophy, theory, and application.* New York: Praeger.

Treffers, A. (1991). Meeting innumeracy at primary school. *Educational Studies*, 22(4), 333–52.

Von Glasersfeld, E. (1982). Subitizing: The role of figural patterns in the development of numerical concepts. *Archives de Psychologie*, 50, 191–218.

Wright, R.J. (1991). The role of counting in children's numerical development. *The Australian Journal of Early Childhood*, 16(2), 43–8.

Wright, R.J., Martland, J. and Stafford, A. (2000). *Early numeracy: Assessment for teaching and intervention.* London: Paul Chapman.

Wright, R.J., Martland, J., Stafford, A. and Stanger, G. (2002). *Teaching number: Advancing children's skills and strategies.* London: Paul Chapman.

Young-Loveridge, J. and Wright, V. (2002). Validation of the NZ Number Framework. In B. Barton, K. Irwin, M. Pfannkuch and M. Thomas (eds), *Mathematics Education in the South Pacific* (pp. 722–9). Sydney: MERGA.

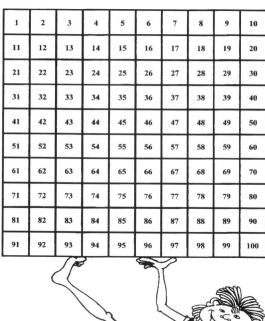

CHAPTER 9

Basic facts and mental computation

In the past, school mathematics programs were dominated by pen-and-paper practice for the four operations of addition, subtraction, multiplication and division. Mental mathematics also featured, with chanting of times tables and 'ten quickies' signalling the commencement of mathematics lessons. In the 1980s, students' understanding of written calculation methods came under scrutiny, with research indicating that flawless written computation performance did not necessarily equate to understanding of the procedures involved (e.g. Carpenter and Moser, 1982; Resnick, 1982). Calls were made for mathematics instruction to ensure that, in promoting computational skill, rich conceptual knowledge was also developed (e.g. Brownell, 1987; Skemp, 1989).

In a technological age, the necessity of teaching standard procedures for written computation is being questioned. In addressing this issue, the National Council for Teachers of Mathematics (NCTM, 1989) stated that instruction in the four operations should provide students with the understanding, skills and confidence to choose the most appropriate method for computation—mental, written or on a calculator. Recently, the importance of mental computation and estimation as the first point for written or calculator methods has been emphasised (White, 2008). In order for these three computational methods to be performed successfully, meaningfully and efficiently, number sense and flexible thinking about numbers are vital.

The Number Framework (Ministry of Education, New Zealand, 2008) outlines nine global stages for learning number that describe flexible thinking strategies characteristic of each stage. The Number Framework stresses that children should not be exposed to standard written algorithms until they use part-whole mental strategies.

Working flexibly with number

Students with good number sense are able to see numbers in a range of combinations and groupings. This knowledge enables them to work flexibly with numbers in mental computational situations. For example, rather than see the addition of 6 and 7 as a fact to be remembered, they see 6 and 4 and 3, which enables them to add the 4 to the 6 to make 10 and then add the remaining 3 to the 10. In another context where 7 is added to 8, the 7 is seen as 2 and 5. In order for students to work flexibly with numbers in this way, they need to have strong number sense and the confidence to work with numbers so they can make use of this knowledge.

Depending on the context, greater or lesser emphasis is placed on developing number sense. In Dutch research and curriculum development, the first five years of schooling focus on developing number sense (over place value) and on promoting the students' intuitive strategies for mental computation. This work has the potential to have a profound impact on how number work is conceptualised.

In most Western classrooms, teaching operations such as 65 + 28 has been undertaken through a place value model. That is, the ones are traded for tens and the answer is duly noted. The Dutch model encourages numbers to be manipulated more flexibly, depending on the context. In this case, the 28 will need to be broken into components that can readily be added to the 65. It is possible that this may be 20 + 5 + 3 so that 20 is added to 65, and then the 5 (to round to 70) and then the remaining 3 is added. In another context, if the 28 were added to 47, then it would make sense to break the 28 into $20 + 3 + 5$. In these cases, 28 is not seen as 2 tens and 8 ones, but rather is broken into components that will make it easy to add to the other number. Students are encouraged to think flexibly with numbers (see Jump and Split strategies discussed in Chapter 8.)

A strategy used to support this flexibility with numbers is that of the empty number line (Beishuizen, 1997). It is a strategy that is progressively built towards using a range of curriculum resources. However, it is a useful strategy for all students when they have good number sense and are able to see numbers as malleable. For example, the number line can assist in tracking and thinking about numbers and operations. Students will use the number line differently depending on their number sense.

The approach suggested here is to provide students with thinking strategies for basic fact calculation that in turn assist in mental computation of larger numbers (two digits and beyond). Research has shown that the main strategy used by poor mental computers is a counting strategy, which often leads to errors (McIntosh and Dole, 2000; Mercer and Miller, 1992; Steinberg, 1985). Research has also shown that explicit teaching of strategies for particular groups of basic facts based on thinking strategies has facilitated fact recall and application in problem solving (e.g. Mercer and Miller, 1992; Steinberg, 1985). In this chapter, instruction is described that aims to promote strong visual images for basic facts through the use of concrete materials. The suggested approach, which is centred around explicit teaching of thinking strategies for particular groups of basic facts, draws from a considerable body of research in the field (e.g. Beishuizen, 1997; Rightsel and Thornton, 1985; Thornton, 1978; Thornton et al., 1983).

The chapter is divided into two sections, with the basic facts of addition discussed first, followed by the basic facts of multiplication. In the past, considerable time was spent on teaching all four operations as if they were not connected. However, research in the Netherlands shows that a greater emphasis on teaching for understanding of the commutativity principle (i.e. $3 + 5 = 5 + 3$) when learning the four operations allows students to develop deeper understandings of operations rather than learning them through rote processes. This chapter reflects such thinking.

Basic facts of addition

The basic facts of addition involve the addition of two addends up to 10. The basic facts of subtraction are the addition facts in reverse. There are 100 basic facts of addition, or 121 if facts with zero are also included. In this section, the facts have been categorised according to a thinking strategy that can assist in understanding and recall of the facts. Similar to the categorisation of the 121 basic addition facts by Thornton et al. (1983), thinking strategies for seven categories of facts are presented here as follows:

1. count-on facts
2. doubles
3. tens facts
4. adding 10 facts
5. bridging 10 (9, 8) facts
6. doubles + 1; + 2 facts
7. last facts.

The 121 facts sorted according to category are displayed in the table, which serves to highlight overlaps of facts between categories. For example, 6 + 4 is in the tens facts category, and also the doubles + 2 category. Similarly, 8 + 3 is a count-on fact as well as a bridging 10 (8) fact. In some cases, particular facts fall within two or more categories and this is apparent as each fact category is presented. The teacher's role in this situation is to encourage discussion of the new fact category and the way in which two different strategies can be used to recall the same fact. Students should be encouraged to express their opinions of particular strategies they might employ in such cases and why.

The table of facts also lists the *spin-around* (the initial fact in reverse order) for each fact in each category. This is to reflect the efficiency of learning facts by strategy and visualisation. Facts are not seen as isolated and individual, but rather as flexible, by simultaneously considering the related spin-around at the time of learning the initial facts. This covertly promotes students' understanding of the commutative principle.

The fact categories are not necessarily hierarchical, but particular groups of facts must be in place before some other fact categories are learnt, as they provide the foundation for deriving other facts. The sequence presented is a suggestion only. Introduction of groups of facts

Basic addition facts sorted by strategy

Count-on facts

Count-on 1	2+1	3+1	4+1	5+1	6+1	7+1	8+1	9+1			
Spin-around	*1+2*	*1+3*	*1+4*	*1+5*	*1+6*	*1+7*	*1+8*	*1+9*			
Count-on 2	3+2	4+2	5+2	6+2	7+2	8+2	9+2				
Spin-around	*2+3*	*2+4*	*2+5*	*2+6*	*2+7*	*2+8*	*2+9*				
Count-on 3	4+3	5+3	6+3	7+3	8+3	9+3					
Spin-around	*3+4*	*3+5*	*3+6*	*3+7*	*3+8*	*3+9*					
Count-on 0	0+1	0+2	0+3	0+4	0+5	0+6	0+7	0+8	0+9	0+10	
Spin-around	*1+0*	*2+0*	*3+0*	*4+0*	*5+0*	*6+0*	*7+0*	*8+0*	*9+0*	*10+0*	
Doubles	1+1	2+2	3+3	4+4	5+5	6+6	7+7	8+8	9+9	10+10	
Tens facts	10+0	9+1	8+2	7+3	6+4	5+5	4+6	3+7	8+2	9+1	0+10
Adding 10	10+0	10+1	10+2	0+3	10+4	10+5	10+6	10+7	10+8	10+9	10+10
Spin-around	*0+10*	*1+10*	*2+10*	*3+10*	*4+10*	*5+10*	*6+10*	*7+10*	*8+10*	*9+10*	
Bridging 10 (9)	9+2	9+3	9+4	9+5	9+6	9+7	9+8				
Spin-around	*2+9*	*3+9*	*4+9*	*5+9*	*6+9*	*7+9*	*8+9*				
Bridging 10 (8)	8+3	8+4	8+5	8+6	8+7						
Spin-around	*3+8*	*4+8*	*5+8*	*6+8*	*7+8*						
Doubles + 1	2+3	3+4	4+5	5+6	6+7	7+8	8+9				
Spin-around	*3+2*	*4+3*	*5+4*	*6+5*	*7+6*	*8+7*	*9+8*				
Doubles + 2	2+4	3+5	4+6	5+7	6+8	7+9					
Spin-around	*4+2*	*5+3*	*6+4*	*7+5*	*8+6*	*9+7*					
Last facts	7+4										
Spin-around	*4+7*										

according to strategy will depend upon the needs of the students and the teacher's own professional judgement.

● Count-on facts

The count-on strategy is where, when given two numbers, one of the numbers is the starting point for counting on the amount of the second number. The count-on strategy is frequently used, but is often the most difficult strategy to overcome once learnt. An alarming research study showed that Year 6 children assessed as having learning difficulties in mathematics could only use primitive counting strategies for mental computation, similar to strategies used by Year 2 *at-risk* children (Ostad, 1998), suggesting that students who experience difficulty with mathematics, and mental computation in particular, have only one strategy upon which to call for mental computation. Many students use the count-on strategy to mentally add two numbers that are far too large to ensure accuracy. In its place, the count-on strategy is very useful for mental computation, but students need to know when and how to use this strategy, and when it is more efficient to use other strategies.

Counting-on 1, 2 and 3

The count-on strategy is only suitable for counting on small numbers such as 1, 2, 3 and zero. In the early years, special instruction needs to be given for counting on zero, and this should be taught after the count-on strategy for numbers 1, 2 and 3 has been consolidated (Booker et al., 1998). When teaching basic fact groups, selection of the examples used needs to be considered carefully so that the usefulness of the strategy is clearly demonstrated. To teach the count-on facts, select examples where one of the two numbers in the pair is considerably greater than 1, 2 or 3. The following facts match this description: $4 + 1, 5 + 1, 6 + 1,$ $7 + 1, 8 + 1, 5 + 2, 6 + 2, 7 + 2, 5 + 3, 6 + 3$, and are the best for initial experience with the strategy.

Visual, auditory and tactile experiences can be used to introduce students to the count-on strategy:

- Using marbles in a tin is a useful teaching approach for developing the count-on strategy (Booker et al., 1998). Show students a group of five marbles, then place them in an empty tin. Let the students see and hear the group of five marbles rolling around in the bottom of the tin. Add one more marble and have the students count on from 5 to 6 as they hear the marble drop into the tin. Practise with other examples. Having the marbles in the tin discourages students from counting the first collection of objects, as they can't see them. The noise created when a marble is added assists in exemplifying the counting-on action.

- Place a group of counters on an overhead projector. Flash the screen to the students and ask them to state how many counters are on the screen. This action capitalises on students' natural ability to *subitise*—that is, recognise the size of a collection of objects without counting. Check by counting this collection if necessary. Cover the collection on the screen and place another two counters on the

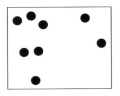

What can you see?

screen and flash the screen again so that only the additional two counters are visible. Ask the students to state how many counters there are altogether. This action discourages the urge to recount the first collection and to count on from this number.

- Flash sets of counters in various count-on fact representations on the overhead projector and ask students to state the solution. Ask the students to explain their strategy. Encourage students to use their subitising skills to determine the amount of counters in each group, then to select the largest group and count on from there.
- Provide students with the fact in symbolic form and ask them to show that fact with counters. Ask students to cover the biggest amount with their hand and to count on the remaining amount.
- As students are presented with particular facts, encourage them to press their fingers on their skin, or to slap their thigh, as they *count on* each number to help keep track of the amount they are counting on.

● The commutative law (spin-arounds or turn-arounds)

The count-on strategy is where the larger of the two numbers is the starting point for counting on the other number. As children explore this strategy through use of concrete materials, they will naturally engage with one of the laws of arithmetic—commutativity: the concept that order doesn't matter. Adding 7 + 2, it is the same as adding 2 + 7. The commutative law must be made explicit to students so they can see that, as they learn each basic fact, they are actually learning *two* facts. They are learning each fact and its *spin-around*.

To explicitly model commutativity, and therefore the spin-around or turn-around notion for the count-on facts, use a piece of paper or card that is divided into two parts. Place counters on each half of the card to model a count-on fact. Ask students to state the fact being modelled. Then physically turn the card around and ask the students to state the fact being modelled. Similarly, as shown in the photo, students can use Unifix to represent the task and twist this in either direction to demonstrate the commutativity principle.

Reinforce the count-on strategy: select the larger of the two numbers and count on the other amount. This strategy can be further reinforced

Student working with spin-arounds—here the student has 4 and 2, or 2 and 4, depending on which way the Unifix blocks are turned

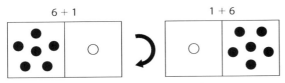

Cards or dominoes reinforce spin-arounds

by providing students with practice sheets of count-on facts. Encourage students to circle the larger of the two numbers as the first step before they count on and then record the answer.

Counting on zero

Special attention needs to be given to counting on with zero. The count-on strategy is just that—counting on: an action is expected and required. However, when counting on zero, no action is required. This may be a potential source of confusion to some students. Through use of concrete materials and careful use of language, students will come to see that counting on with zero results in nothing being added to the situation and the resulting answer is the original amount started with. Commutativity is reinforced through the strategy for counting on where the larger of the two numbers is selected as the starting point.

● Doubles facts

The doubles are usually the most easily learnt facts for students. Often students know their doubles facts up to 5 + 5 before they come to school. Visualising real objects and 'things' can assist students to learn doubles (Thornton, 1982). When introducing the doubles thinking strategy, encourage students to brainstorm *things* that *come in* a particular amount that can be doubled. For example, the number of wheels on a bicycle is two—bicycle wheels *come in* twos. So two bicycles is four wheels (double two is four). The following is a list of suggestions that can be used to help children visualise a particular number and then double it:

1—person, stop sign, sun
2—drumsticks, eyes, ears, bicycle wheels
3—clover leaf, cricket stumps, tricycle wheels
4—car tyres, legs on a dog/cat/horse, etc.
5—fingers
6—legs on an insect
7—calendar (2 weeks)
8—legs on an octopus/spider
9—Channel 9 TV symbol.

When teaching doubles, focus on the language and a visual image—'5 fingers on one hand; 10 fingers on 2 hands; 5 and 5 is 10; double 5 is 10'. When students falter in thinking of the solution to 8 + 8, have them think of a spider: 'How many legs does a spider have? Now double it.' For 7 + 7 think of a calendar: 'How many days is 2 weeks?' For 6 + 6: 'How many legs do 2 insects have?' The images students use to assist in learning the doubles should come from them. They need to select the item that most appeals to them and then double it. (Too many images can be a source of confusion, however.)

Doubles plus one and doubles plus two

Once the doubles are mastered, they can be used as a platform for facts that are almost doubles. Facts that fall into this category are the doubles plus one facts (2 + 3, 3 + 4, 4 + 5, 5 + 6, 6 + 7, 7 + 8, 8 + 9) and the doubles plus two facts (2 + 4, 3 + 5, 4 + 6, 5 + 7, 6 + 8, 7 + 9). As a category, these facts should be introduced once the doubles facts have been consolidated. Recognising doubles plus one and doubles plus two facts requires considerable cognitive effort, so it should be delayed until students are familiar with easier categories of facts.

As students learn more and more facts, particular facts may previously have been learnt through another strategy. This negates the need to re-learn the same fact with a different and possibly more difficult strategy. Through examination of the list of doubles plus one and doubles plus two facts, it can be seen that some facts fit within other fact categories. The following facts have been covered through the count-on strategy: 2 + 3, 2 + 4, 3 + 4, 3 + 5. The 4 + 6 fact is a simple tens fact and is discussed below. The facts 7 + 8, 8 + 9 and 7 + 9 belong to the bridging 10 facts category and may be simpler for some students to learn through that strategy than as an almost double fact. The almost double facts have been presented here because of their link to the doubles facts.

To assist students to visualise facts that are almost doubles, use counters on 2 cm grids. Use different coloured counters to show the two numbers in the fact. Display counters as a mirror image with an extra counter clearly visible for doubles plus one facts and two extra counters for doubles plus two facts. When the counters are displayed, describe the fact as being a particular double plus one or plus two. From the examples given below, it can be seen that 5 + 6 is the same as double 5 plus 1 more. Similarly, 6 + 8 is the same as double 6 plus 2 more.

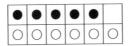

5 + 6 is the same as double 5 + 1

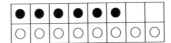

6 + 8 is the same as double 6 + 2

For some students, these facts may be more appropriately recognised as doubles *less* 1 or 2. That is, 5 + 6 may be seen more easily as double 6 less 1, 6 + 8 may be double 8 less 2. Promote this type of discussion around the visual images provoked by the concrete representations, and encourage students to use the strategy that most appeals to them. It is better if students select and use one strategy until particular facts become consolidated, otherwise confusion may occur.

● Tens facts

The tens facts are those facts where the number pair combines to give a total of 10. Recognising combinations to 10 is a frequently used strategy in mental computation. Tens facts are often readily acquired, but it is important to ensure that all tens facts are equally familiar to students so that they have access to a valuable strategy for mental computation.

Orientation to the tens facts is best achieved through use of a tens frame. A tens frame is a 2 × 5 grid with grid cells large enough to accommodate a coloured counter. Before using the tens frame to represent tens fact combinations, students need to be able to instantly recognise numbers from 0 to 10 as represented with counters on the grid. This is promoted through positioning counters in a particular consistent arrangement (Dole and Beswick, 2001) and through drawing attention to the number of counters, and the number of counters missing, as follows for the numbers 10, 9, 8, 5, 6 and 7:

10—no counters missing

9—one missing

8—two missing

5—top line filled

6—five and one more

7—five and two more.

The numbers 0 to 4 are readily recognised by students. The arrangement of counters on the ten frame for the numbers zero to tens are presented in the diagram.

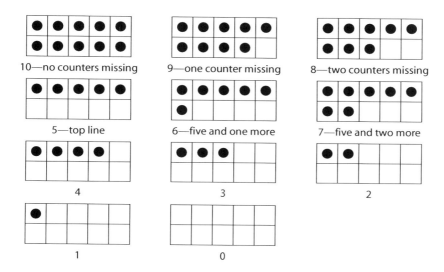

Once students can readily recognise numbers 0 to 10 on the tens frame without counting, link tens fact pairs by emphasising the amount needed to fill the tens frame for the following five facts: 10 and 0, 9 and 1, 8 and 2, 5 and 5, 7 and 3, 6 and 4. Once these five number combinations have been consolidated, the rest of the tens facts are the spin-arounds.

The sequence for presentation of the tens facts and accompanying language is as follows:

$10 + \square = 10$ What goes with 10 to give 10?

$9 + \square = 10$ What goes with 9 to give 10?

$8 + \square = 10$ What goes with 8 to give 10?

$5 + \square = 10$ What goes with 5 to give 10?

$6 + \square = 10$ What goes with 6 to give 10?

$7 + \square = 10$ What goes with 7 to give 10?

$0 + \square = 10$ Zero and what gives 10?

$1 + \square = 10$ One and what gives 10?

$2 + \square = 10$ Two and what gives 10?

$3 + \square = 10$ Three and what gives 10?

$4 + \square = 10$ Four and what gives 10?

Adding tens facts

Understanding what happens when a ten is added to a number relates to place value and numeration knowledge. To explore this notion with students, counting activities with a calculator are useful. If a calculator has an inbuilt constant function, it can be programmed to count by entering a starting number, pressing the addition key and then entering the number by which you wish to count. Try the teaching ideas in the box. If neither of these methods works (because the calculator does not have the constant function), try another calculator.

T E A C H I N G   I D E A S

Adding tens facts

To make your calculator count in tens:
Enter any number (e.g. 6),
Press the add key, and
Then enter 10, and
Press the equals button repeatedly.
Observe the changes occurring . . .

Alternatively:
If that doesn't work, try entering the starting number, then press the add key twice before entering the counting number.
Then press equals repeatedly.

Take the example of counting in tens from 6. The idea is to make students focus on what is happening every time the equals button is pressed. Students will see that when adding ten, the number in the ones place remains constant, but the digits to the left of the entered number increase by one every time. The pattern for 6 + 10 repeatedly is 16, 26, 36, 46, 56, 66 . . . and so on. Try other starting numbers and see which digits change and which stay the same. Link this to a general principle of adding ten to a number—the digit in the ones place stays the same but a ten is added to the digit in the tens place. From this generalisation, students can quickly determine the solution to facts in which ten has been added. This is an extremely useful strategy for mental computation of numbers with more than two digits.

Bridging ten facts

The bridging ten facts are when one number of the fact pair is close to 10. Bridging ten is very useful when one digit is 9 (9 + 4, 9 + 5, 9 + 6, 9 + 7, 9 + 8) or even 8 (8 + 4, 8 + 5, 8 + 6, 8 + 7), but for numbers smaller than that the strategy loses its usefulness and efficiency. To show students how to use the bridging ten strategy and to provide them with a strong visual image of this strategy in action, use a double tens frame.

A double tens frame consists of two tens frames displayed one above the other. In the top frame place nine (or eight) counters in one colour. (Note that 'filling' of ten frames is consistent with that suggested for developing tens facts.) On the bottom frame, display the second number in the fact pair with counters in a different colour. The top frame represents the number in relation to 10. Clearly, when the number is 9, one more counter is required to fill the ten frame. For 8, two more counters are required. Take the required number of counters from the bottom frame to fill the top frame. The thinking behind this strategy is to think of the first number as 10 and alter the second number accordingly. For example, 9 + 6 is 10 + 5; 8 + 5 is 10 + 3. The arrangement of counters on the tens frames for the facts 9 + 4 and 8 + 5 are presented in the diagram.

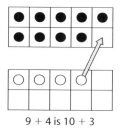

9 + 4 is 10 + 3

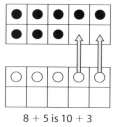

8 + 5 is 10 + 3

Double tens frame to show the thinking behind the bridging ten strategy

● Last facts

By highlighting facts on an addition grid as each fact category is introduced, all of the 121 facts are covered by the preceding strategies, except one. That fact is 4 + 7, or 7 + 4, the only one which does not fit simply into any category. In this light, this fact is special and should be indicated to students as such. With this fact, strategies for solution are a discussion point. Some students may see it as an almost ten fact, some students may see it as a bridging ten fact, other students may simply see it as a special fact that gives 11.

Basic facts of subtraction

Once all addition facts have been consolidated, subtraction facts should easily follow. Each fact should be considered as a set of four facts, two addition and two subtraction facts, generated through rearrangement of the three numbers in the fact. For each fact encountered, students learn two facts using the commutative law—for example, 4 + 6 = 10; therefore 6 + 4 = 10. Two subtraction facts are generated from the addition facts: 10 – 6 = 4, 10 – 4 = 6. Within each fact, the three numbers share a special relationship.

The strategy for subtraction facts is to *think addition*. The think addition strategy is best applied once the addition facts have been fully consolidated—that is, when students can rapidly determine the solution to the addition fact. When they are presented with a subtraction fact, attention is drawn to the related and previously learnt addition fact and the number combinations within that fact. For example, when presented with 10 – 6, to think addition is to think of the number that, when combined with 6, makes 10.

Addition and subtraction: Mental computation

95 + 53 …
30 + 146 …
75 + 38 …
84 + 87…
78 — 33…

How would you work these out?

Having ready recall of basic facts assists in mental computation. The strategies outlined in the previous section for promoting recall of basic addition facts were presented as ways in which learning basic facts can be facilitated through provision of visual images and thinking strategies. The thinking strategies are also useful for mental computation. Consider how you would mentally compute the examples in the diagram.

Common strategies for mentally computing solutions to these examples will include the strategies outlined for learning the basic facts. In the first example, it is possible to see that the bridging ten strategy can be used when adding 90 and 50; in the second example, the adding ten strategy can be used; for the third example, knowledge of tens facts is useful; and in the fourth example, doubles are presented. For the last example, another strategy may be useful—the count-back strategy for subtraction. Like the count-on strategy, this strategy is useful for counting back 1, 2 or 3, but becomes inefficient and prone to error when used to count back larger amounts.

The suggested methods for mental computation are not the only ways in which solutions may be determined, and mental methods are diverse and often quite idiosyncratic (McIntosh and Dole, 2000). The important thing is that learners are given support and encouragement to determine their own solution methods, and that they have the confidence to do so. Armed with automatic recall of basic addition and subtraction facts, as well as thinking strategies for computation, learners are in a strong position to invent their own procedures for computation which will be inherently meaningful to them.

Basic facts of multiplication

Automatic recall of basic multiplication facts is an important objective of primary school mathematics, and is vital for mental computation and estimation. There is no question about the value of students learning these basic facts, but the approach taken must be considered. Why students need to learn their tables beyond the ten times also must be questioned. For efficient mental computation, students need automatic recall of all multiplication facts up to ten times. From this, they can derive solutions to eleven and twelve times tables and beyond.

● Foundations for basic multiplication facts

Counting patterns lay the foundation for multiplication facts. Skip counting in various amounts encourages students to look for patterns in counting, and this builds number sense. Early skip counting activities should provide students with some form of concrete reference so that the size of numbers being counted can be conceptualised. The following are some ideas for early counting in tens, fives and twos where the students are standing in a line and progressively count in a specified amount.

- For counting in tens, the first student in line brings his or her hands forward, stretching out the fingers while shouting 'ten'. The second child in line does the same but shouts 'twenty'. This continues down the line until all students have raised their hands.

- For counting in fives, the first student brings one hand forward and shouts 'five', then the other hand forward and shouts 'ten'; the second student brings one hand forward and shouts 'fifteen', then 'twenty', and so on down the line.
- For counting in twos, the students stand in line and at the count of 'two', 'four', 'six', etc., each student in turn blinks (i.e. using two eyes), with a nod of the head.

Follow-up activities for these counting sequences can include making posters using cut-outs of students' hands, and cutting out magazine pictures of eyes that are displayed in a linear fashion with the counting sequence displayed underneath.

Displays of counting sequences enable patterns to be determined. When counting in fives, the numbers in the sequence end in 5 or zero; when counting in tens, each counted number ends in zero; and when counting in twos, the sequence is 2, 4, 6, 8, 0 before the sequence repeats again, starting with 2. From previous exploration of odd and even numbers, students will recognise that the pattern of counting in twos is also a way of identifying even numbers. Understanding of number relationships and patterns in number builds number sense.

● Counting patterns and the one-hundred board

Displaying counters on a one-hundred board is useful for skip counting and patterning. On a one-hundred board, the numbers 1 to 100 are displayed in a 10 × 10 array, with the numbers 1 to 10 displayed along the first row, and the numbers 91 to 100 displayed along the bottom row. The one-hundred board is actually a number line represented as an array rather than in a linear representation.

Select a number between 2 and 9 and place a transparent counter or block on each multiple to create a visual representation of the counting sequence. Focus students' attention on the digits in the ones place in the sequence, noting the pattern of digits and when the pattern recurs. For counting in nines, the one-hundred board emphasises how counting in nines is one less than counting in tens as the placement of counters on the board shows a diagonal pattern. Key ideas associated with counting in nines emerge through analysing this arrangement of the last digit of numbers in the sequence and noting that they follow a pattern of counting backwards in ones. Also, the digits in each number in the nines sequence total 9 (18 is 1 + 8 which equals 9; 27 is 2 + 7 which equals 9; 36 is 3 + 6 which equals 9, and so on). However, some counting patterns are quite complex and the one-hundred board display may add to their complexity. For some students, the one-hundred board may not be a useful representation. And clearly, not all patterns will be explored on the one-hundred board at once, as this would lead to information overload. The one-hundred board is only one way to explore patterns.

Students working on the one-hundred board

Sequencing basic facts

Certain groups of multiplication facts are easier to learn than others. The following is a suggested sequence for instruction (based on suggestions from Baturo and English, 1982) whereby simpler facts are consolidated in order to reduce the number of individual facts to be learnt. For example, students find the twos, fives and tens facts relatively easy to recall. When these facts are learnt, students can readily access 44 of the 121 multiplication facts. In the following section, ideas for promoting understanding and recall of the basic facts of multiplication are described in the following order:

- twos, fives and tens facts
- ones and zeros facts
- pattern facts
- square numbers
- threes and sixes facts, and
- last facts.

● Twos, fives and tens facts

The easiest multiplication facts are the twos, fives and tens. Experience with skip counting in these amounts lays the foundation for learning the two, five and ten times tables. The first ten numbers in each counting sequence need to be signalled to students as being of particular interest for learning the multiplication facts. Although the easiest facts, each set of facts should be treated separately until that set is consolidated.

Twos facts (doubles)

The twos facts should be the first facts that students learn in the formal study of multiplication facts. From the study of addition facts, students should already know their two times tables, as these facts are actually the addition doubles. Linking the twos facts to addition doubles connects multiplication to repeated addition. The symbolic representation of the addition doubles fact and counters in appropriate arrays should be presented with the multiplication fact in symbolic form so that connections can be seen.

$$6 + 6 = 12 \qquad \qquad 6 \times 2 = 12$$

A group of 6, times 2

As each double is related to multiplication, its associated spin-around fact is also presented. The answer to 6×2 is the same as 2×6.

Tens facts

Counting in tens has a simple rhythm (ten, twenty, thirty . . .) and is easily learnt by students. Making bundles of ten icy-pole sticks secured with rubber bands provides a physical referent for counting in tens and also builds number sense in terms of conceptualising the size of numbers. As each group of ten is added, the number of groups and the total amount need to be consciously recognised. If there are three groups of ten, there are 30 icy-pole sticks. A quick check of tens facts will reveal which particular students in your class need further support for their tens facts.

Fives facts

Early explorations of counting in fives reveal that all numbers in the sequence end in 5 or zero, and this is a check when learning the five times tables. An analogue clock can also be used as a reference point, in that the numbers 1 to 12 on the clock are related to five-minute intervals. When multiplying by five, the number of fives to be calculated can be located on the clock and counting in fives can occur from the one to that point to give the answer. For example, 7×5, locate the 7 on the clock and skip count in fives to reach 7 (35).

● Ones and zeros facts

Multiplying by one and multiplying by zero often cause difficulty for students, as they confuse the solution with that which occurs when adding one or zero. Special attention needs to be given to these two groups of multiplication facts so that students have something to which they can refer when they are feeling confusion as they recall the one and zero times tables.

Ones facts

As with previous facts, the interpretation of the symbolic representation with the concrete representation provides a mental clue for checking answers. A group of counters is displayed and discussed as one group of a particular number of counters. The matching written form is given and analysed. If there are seven counters, this is described as 'a group of seven taken one time'. The answer is 7. From this, the spin-around fact is derived so that 7×1 is the same as 1×7. The generalisation is that, upon multiplying by one, the solution is the starting number. This can be confirmed with a calculator. A real-world situation that can be invoked when multiplying a number by one is to think of queues. Students line up at the tuckshop and know their place in the line. If there are eight others ahead of them in the line, then they are modelling 9×1, which equals 9.

Zeros facts

Multiplying by zero is difficult in terms of finding a real situation in which to contextualise why multiplication with zero results in zero. Using a calculator can support students' knowledge that a number multiplied by zero is zero, or zero multiplied by a number is zero. Patterning may also help. By presenting multiplication equations in a particular sequence, students can predict that the next number in the sequence is zero:

$$5 \times 3 = 15$$
$$5 \times 2 = 10$$
$$5 \times 1 = 5$$
$$5 \times 0 = 0$$

Division by zero

One of the tricky facts to learn is division by zero. When students enter the following in a calculator, they may be confronted by an unexpected result:

$$0 \div 12 = 0$$
$$12 \div 0 = \text{error}$$

Patterning can assist students to make meaning of the second situation. By observing the following sequence, it can be seen that as you divide by increasingly smaller amounts, the solution gets bigger and bigger:

$$12 \div 12 = 1$$
$$12 \div 6 = 2$$
$$12 \div 4 = 3$$
$$12 \div 2 = 6$$
$$12 \div 1 = 12$$
$$12 \div 0.5 = 24$$
$$12 \div 0.1 = 120$$
$$12 \div 0.01 = 1200$$
$$12 \div 0.001 = 12\ 000$$

When dividing by a very small number—a number that gets closer and closer to zero, the solution becomes very large. We can only imagine the size of the solution if we actually did divide by zero. The solution is undefined.

● Patterns facts: Fours, nines, square numbers

Fours facts

The fours facts require facility with the twos facts because the strategy is to think of doubling the twos. Visually, the twos facts are shown and then the fours facts shown as a mirror image of the twos facts. The diagram shows 7×4, where this fact is considered as 7×2 (which equals 14) and is then doubled (to give 28).

7 × 4 is 7 × 2 (14) then doubled (28)

Nines facts

Learning the nines facts is a process of using two thinking strategies. The first is to relate the fact to known tens facts to give an estimate of the correct answer. For example, to determine 9×6, find 10×6 (60). Because 9 is less than 10, the solution will be 50-something. To determine what the correct solution will be is to know the patterns made when multiplying by nine, as the answer to each fact gives a combination of digits that add to nine:

$9 \times 2 = 18$	$1 + 8 = 9$
$9 \times 3 = 27$	$2 + 7 = 9$
$9 \times 4 = 36$	$3 + 6 = 9$
$9 \times 5 = 45$	$4 + 5 = 9$
$9 \times 6 = 54$	$5 + 4 = 9$
$9 \times 7 = 63$	$6 + 3 = 9$
$9 \times 8 = 72$	$7 + 2 = 9$
$9 \times 9 = 81$	$8 + 1 = 9$

9 is nearly 10
$10 \times 6 = 60$
$9 \times 6 = 50$-something

As the solution to 9×6 is 50-something, the answer must be 54, as $5 + 4 = 9$ and this fits the pattern of multiplying with nines. Prior experience with skip counting and exploration of the representation of counters on the one-hundred board will provide further confirmation of the accuracy of the solution.

Square numbers

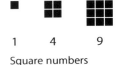

1   4   9

Square numbers

Many students do not realise that square numbers are given their name because that is the shape that the numbers make when they are represented using blocks or counters. It is important for students to know the square numbers as a group of special numbers, as square number patterns are frequently encountered in algebra and other number study. Early exploration with square numbers should provide students with the opportunity to discover the first ten square numbers themselves through the use of concrete materials.

When linking square numbers to the multiplication facts, the image of blocks or counters in square arrays should be a visual referent point that students can use to assist in solution recall.

● Remaining facts

After learning the twos, fives, tens, ones, zeros, fours, nines and square number facts, almost all of the 121 multiplication facts have been covered, as seen on the grid. The remaining facts are indicated by the grey cells. There are 12 empty cells in the grid, but only six facts remain to be learnt as the other six are spinarounds. There are three facts from the threes (3×6; 3×7; 3×8), two facts from the sixes (6×7; 6×8) and one from the sevens (7×8).

Threes facts

One way to help students learn the remaining threes facts is to encourage the use of a build-up strategy, where the fact is linked to the known twos fact. For example, the 3×8 fact can be considered in terms of 2×8, or doubling 8, and then adding another set of 8 to that double. The visual representation can also provide a mental picture to assist in the thinking process.

x	0	1	2	3	4	5	6	7	8	9	10
0	0	0	0	0	0	0	0	0	0	0	0
1	0	1	2	3	4	5	6	7	8	9	10
2	0	2	4	6	8	10	12	14	16	18	20
3	0	3	6	9	12	15				27	30
4	0	4	8	12	16	20	24	28	32	36	40
5	0	5	10	15	20	25	30	35	40	45	50
6	0	6	12		24	30	36			54	60
7	0	7	14		28	35		49		63	70
8	0	8	16		32	40			64	72	80
9	0	9	18	27	36	45	54	63	72	81	90
10	0	10	20	30	40	50	60	70	80	90	100

Remaining facts

Sixes facts

A similar strategy can be used for learning the remaining sixes facts, where the related fives fact is used as the starting point. For example, to find 6×8, think 5×8 (40) and then add another set of 8 (48).

$3 \times 8 = (2 \times 8) + (1 \times 8)$

The last fact

The only fact not covered by any of these strategies is 7×8 (and its spin-around, 8×7). For this fact, students have a variety of thinking strategies at their disposal, acquired through learning all previous facts. This fact can be solved using any number of strategies, but as it is the only fact in this category, the students should feel free to use whatever method they prefer and to discuss their choice.

3×8 is 2×8 (double $8 = 16$) and then another set

Basic facts of division

As with deferring subtraction facts until addition facts are known, the division facts should be deferred until the multiplication facts are consolidated. Learning division facts is assisted by automatic recall of multiplication facts and number sense, in terms of understanding the relationship between the three numbers in each multiplication fact.

Each multiplication fact consists of two factors, which are multiplied together to give a third number, the product. The structure of multiplication facts is *factor × factor = product*. Rearrangement of the three numbers in each multiplication fact gives two corresponding division facts—for example:

$4 \times 7 = 28$ $28 \div 4 = 7$
$7 \times 4 = 28$ $28 \div 7 = 4$

The strategy for division facts is to *think multiplication*. When provided with a division problem, students should think of what number, when multiplied by the given answer, equals the starting amount. The structure of division facts is *product* divided by a *factor* to give the other *factor* (product ÷ factor = factor).

Mental computation and estimation

Automatic recall of the 121 basic multiplication facts assists mental computation and estimation. Consider the following exercises:

7×50 5×25

5×19 4×26

10×40 6×49

3×99 7×104

With ready recall of basic facts and with number sense that incorporates place value, mentally determining the solution to the above exercises should be a relatively straightforward exercise. Addition strategies are also seen to play a role in attaining a solution. For example, to calculate 5×19, a simple path may be to round 19 to 20, multiply 20 by 5 and then take away the 5 ones that were added when 19 was rounded to 20.

Providing students with opportunities and encouragement to practise mental computation with multiplication and division helps them come to rely on their own mental ability to calculate. With confidence in mental computation ability, students are in a position to choose the most appropriate computational method in any given situation—pen-and-paper, mental or calculator.

● Estimation

Estimation assists in all computation situations involving written, mental or calculator methods. Having an estimate of the answer provides a checking mechanism against which the calculated answer can be compared and the reasonableness of the result determined. Although very useful for computation, estimating is not a natural process, and students need to be taught specific strategies for estimating.

Frequently, the processes of estimation and mental computation are considered to be the same. However, this is not the case. Estimating the cost of two music CDs at $27 each means coming up with an approximate answer, possibly by rounding $27 to $30 and then multiplying by two to give $60. Mental computation is to find the exact answer. The estimated answer is around $60; the exact answer is $54.

Front-end estimation

One of the skills of estimation is called *front-end estimation*, where the first digits in each number are the point of focus. For example, to estimate 56 × 74, the front digits of 50 and 70 are considered and the multiplication is carried out. When estimating the sum of three numbers such as 43, 49 and 46, the front fours in each number are added together (or one of the fours is multiplied by three) to give an estimate of the solution. Front-end estimation is useful as a beginning point for estimating answers, but often further estimation needs to be conducted for a better estimate to be made.

● Compatible numbers

Another strategy for estimation is finding compatible numbers. For example, to estimate 25 × 28 × 4, a beginning point would be to multiply 25 × 4, as this yields 100. These two numbers are compatible. The ability to see compatible numbers is a product of number sense and having a wide-ranging repertoire of knowledge about number relationships.

Rounding

Rounding numbers can also assist in seeing compatible numbers. In the following examples, the numbers in each exercise have been rounded to be compatible. Once the numbers are compatible, estimation becomes a much simpler task:

38 + 67 + 49 + 56	becomes	35 + 65 + 50 + 50
64 × 8	becomes	60 × 8
60 divided by 8	becomes	64 divided by 8
2637 divided by 4	becomes	2800 divided by 4

Good estimates and better estimates

Often students find rounding and estimating to be a tedious process, one that is usually required of them before they are asked to complete written calculation procedures. Because students often find written calculation procedures to be tedious in themselves, they may perceive rounding and estimating to make the task even longer and more tedious.

To encourage students to develop and use estimation skills, they need to be convinced that such skills are valuable. This frequently means finding alternative reasons for estimating and rounding that are not extra tasks when practising written procedures.

Estimation is a process of finding good estimates and then making better estimates. Students need to be engaged in discussion as to when it is appropriate to over-compensate or under-compensate as they make estimations.

● The daily mentals block

A structured, daily focus on mental computation will promote students' facility with basic facts and build number sense. The first ten minutes of every mathematics lesson should be devoted to mental mathematics activities and exercises. In the past, mental mathematics consisted of speed and practice exercises, with students chanting the basic facts of multiplication and then being tested on their recall of the facts through timed tests. Often punitive action fell on those students who did not know their 'tables'. However, a new approach to mentals and the ten 'quickies' can benefit students when implemented with care and consideration of students' self-concept.

Sets of ten quickies can be used to settle the class at the commencement of the lesson, and if little attention is drawn to those who do not get 10 out of 10 correct, students will continue to put in effort and energy to enhance their mentals score. The ten quickies are a great way to warm up the brain. Each set of ten quickies should be previously learnt maths facts.

After a few sets of ten quickies have been used as a warm-up, introduce the targeted strategy of focus, and this should be the focus for the week. As instant recall of basic facts is the foundation for mental computation, it is important that you support students to acquire these facts. Provide opportunities for students to explore the new strategy, discuss it and share similar strategies. Students should then be provided with varied practice activities so that they can apply the strategy and reinforce the basic facts to which the strategy applies. A test at the end of the week can provide students with feedback on progress. Once a fact category has been targeted, this can be included in future sets of ten quickies used at the start of the lesson. The aim is for automatic recall of basic addition and subtraction facts to support mental computation, estimation and number sense activities. Once mastery of basic addition and multiplication facts has been attained, build other mental computation activities into the daily mentals block (e.g. fractions, percentages, decimals, two-digit addition and subtraction, measurement conversions, shape names and properties).

Promoting mental computation in the middle years

Mental computation provides fluency in number and flexibility in using and applying number knowledge in various situations. In the middle years, a structured mental computation program is an essential component of the yearly mathematics curriculum. Research has shown that poor mental computers in the middle years rely on primitive counting strategies that are prone to error (Ruthven, 1998). In a classroom where mental computation is an integral part of a mathematics lesson, where students can share their thinking strategies and can learn new strategies and approaches without ridicule, students' number knowledge and confidence in learning will flourish. In this chapter, a 'strategies approach' for teaching the basic facts was outlined. Through such an approach, students develop strategies for thinking about the basic facts that support recall of the facts when that recall fails. The aim of a basic facts program is for automatic recall, but the pathway to this is not through rote drill and practice. The middle years teacher must diagnose students' difficulties with basic fact recall and target those groups of facts that are not part of students' automatic repertoire. After basic fact mastery, two-digit addition and subtraction exercises should follow. After that, target multiplication and division using single-digit numbers multiplied or divided by two-digit numbers. Every new number topic in the middle years curriculum is an opportunity to promote mental computation. When doing fractions, promote addition and subtraction where mental images assist students to solve the exercises:

$2 - \frac{5}{6}$

$\frac{3}{4} + \frac{6}{8}$

$\frac{5}{8} + \frac{5}{8}$

Ask students to visualise what is happening when you are trying to determine how many halves in 3. When they say that they can see three pizzas divided into two pieces each and they tell you that they can see six, show them that they have just calculated 3 divided by ½. Ask students to think of a number line and determine how many 0.2s in one whole. Then ask them to mentally calculate 3 divided by 0.2 by asking them how many 0.2s are in 3. By posing fraction and decimal division questions in language that supports meaning, students' ability for mental computation of these problem types will be enhanced. For multiplication, think of groups. For example, $6 \times ½$ can be regarded as six groups of ½. This means that the solution is 3. When studying percentages, provide students with opportunities to mentally calculate percentages of given amounts, but carefully sequence the exercises. Begin with asking students to find 50 per cent of given amounts, linking 50 per cent with ½. Then extend this to finding 25 per cent, where students find half and then half again. But also discuss other strategies. For example, students may find it easier to find one-quarter of an amount when the amount is readily recognised as being a multiple of 4 (e.g. 25 per cent of 800). The halving and halving strategy may work better for other numbers (e.g. 25 per cent of 968). Include mental computation exercises when studying measurement topics. Provide students with opportunities to convert between units of measure. For example, when studying length, reinforce the place value nature of SI metric units, by providing students with practice in changing centimetres to metres (times by 100), metres to centimetres (divide by 100), centimetres to millimetres (divide by 10) and so on. As outlined above, daily mental computation practice is an importance element of a mathematics lesson, and this is extremely important in the middle years.

■ REVIEW QUESTIONS

9.1 Provide an argument for or against explicitly teaching subtraction facts.

9.2 Describe the value of the count-on strategy, and also the danger of this strategy, for students.

9.3 What is the value of teaching basic facts? Provide some specific examples to support your answer.

9.4 Why is number sense so important for learning basic multiplication facts? In your response, provide a definition of number sense and provide examples of how number sense might be used to assist in developing multiplication facts.

9.5 Outline a teaching plan for developing the basic facts of addition or the basic facts of multiplication.

Further reading

Beishuizen, M. and Anghileri, J. (1998). Which mental strategies in the early number curriculum? A comparison of British ideas and Dutch views. *British Education Research Journal*, 24, 519–38.

Bobis, J. F. (1991). Using a calculator to develop number sense. *Arithmetic Teacher*, 38(5), 42–5.

Brinkworth, P. (1998). How do you subtract? *Australian Primary Mathematics Classroom*, 3(3), 8–10.

Harte, S. W. and Glover, M. J. (1993). Estimation is mathematical thinking. *Arithmetic Teacher*, 41(2), 75–7.

Huinker, D. (2002). Calculators as learning tools for young children's exploration of numbers. *Teaching Children Mathematics*, 8(6), 316–21.

Isaacs, A. and Carroll, W. (1999). Strategies for basic facts instruction. *Teaching Children Mathematics*, 5(9), 508–15.

Leutzinger, L. (1999). Developing thinking strategies for addition facts. *Teaching Children Mathematics*, 6(1), 14–18.

Morgan, G. (2000). Put mental computation first? *Australian Primary Mathematics Classroom*, 5(3), 4–9.

Randolph, T. and Sherman, H. (2001). Alternative algorithms: Increasing options, reducing errors. *Teaching Children Mathematics*, 7(8), 480–4.

Sowder, J. and Schappelle, B. (1994). Number sense-making. *Arithmetic Teacher*, 41(6), 342–5.

Thompson, A. and Sroule, S. (2000). Deciding when to use calculators. *Mathematics Teaching in the Middle School*, 6(2), 126–9.

References

Baturo, A. and English, L. (1982). *Sunshine Mathematics 7*. Melbourne: Leyman Cheshire.

Beishuizen, M. (1997). Mental Arithmetic: Mental recall or mental strategies? *Mathematics Teaching*, 160, 16–19.

Booker, G., Bond, D., Briggs, J. and Davey, G. (1998). *Teaching primary mathematics.* Sydney: Pearson.

Brownell, W.A. (1987[1956]). AT classic: Meaning and skill—maintaining the balance. *Arithmetic Teacher,* 34(8), 18–25.

Carpenter, T.P. and Moser, J.M. (1982). The development of addition and subtraction problem solving skills. In T.P. Carpenter, J.M.Moser and T.A. Romberg (eds), *Addition and subtraction: A cognitive perspective.* Hillsdale, NJ: Lawrence Erlbaum.

Dole, S. and Beswick, K. (2001). Developing tens facts with grade prep children: A teaching experiment. In J. Bobis, B. Perry and M. Mitchelmore (eds), *Numeracy and beyond: Proceedings of the 24th annual conference of the Mathematics Education Research Group of Australasia* (pp. 186–93). Sydney: MERGA.

McIntosh, A. and Dole, S. (2000). Mental computation, number sense and general mathematics ability: Are they linked? In J. Bana and A. Chapman (eds), *Proceedings of the 23rd Annual Conference of the Mathematics Education Research Group of Australasia* (Vol. 2, pp. 401–8). Fremantle: MERGA.

Mercer, C. and Miller, S. (1992). Teaching students with learning problems in math to acquire understanding and apply basic math facts. *Remedial and Special Education,* 13(3), 19–35, 62.

Ministry of Education, New Zealand. (2008). *The Number Framework.* Wellington: Ministry of Education.

National Council of Teachers of Mathematics (1989). *Curriculum and evaluation standards for school mathematics.* Reston, VA: NCTM.

Ostad, S.A. (1998). Developmental differences in solving simple arithmetic word problems and simple number-fact problems: A comparison of mathematically normal and mathematically disabled children. *Mathematical Cognition,* 4(1), 1–19.

Resnick, L. (1982). Syntax and semantics in learning to subtract. In T.P. Carpenter, J.M. Moser and T.A. Romberg (eds), *Addition and subtraction: A cognitive perspective.* Hillsdale, NJ: Lawrence Erlbaum.

Rightsel, P.S. and Thornton, C.A. (1985). 72 addition facts can be mastered by mid-grade 1. *Arithmetic Teacher,* 33, 8–10.

Ruthven, K. (1998). The use of mental, written, and calculator strategies of numerical computation by upper primary pupils within a 'calculator-aware' number curriculum. *British Educational Research Journal,* 24 (1), 21–42.

Skemp, R. (1989). *Prologue: Relational understanding and instrumental understanding.* London: Routledge.

Steinberg, R. (1985). Instruction on derived facts strategies in addition and subtraction. *Journal for Research in Mathematics Education,* 16(5), 337–55.

Thornton, C.A. (1978). Emphasising thinking strategies in basic fact instruction. *Journal for Research in Mathematics Education,* 9, 214–27.

——(1982). Doubles up—easy! *Arithmetic Teacher,* 29, 20.

Thornton, C.A., Jones, G.A. and Toohey, M.A. (1983). A multisensory approach to thinking strategies for remedial instruction in basic addition facts. *Journal for Research in Mathematics Education, 14*, 198–203.

White, A. (2008). Learning mathematics in the middle years. In H. Forgasz, A. Barkatsas, A. Bishop, B. Clarke, S. Keast, W. Seah and P. Sullivan (eds), *Mathematics Education Research Group of Australasia: Research in Mathematics Education in Australia 2004–2007* (pp. 17–40). Rotterdam: Sense Publishers.

CHAPTER 10

Written algorithms

A n algorithm is a prescribed sequence of steps that, when executed correctly, results in the desired outcome. Real-world examples of algorithms include using an automatic teller machine, building a model plane, putting together a trampoline, and even possibly making a cup of coffee. When the steps are executed in these algorithms in the prescribed manner, the results are favourable. If particular steps are not executed as prescribed, the result may be less than satisfactory. In mathematics, algorithms are procedures for performing calculations to achieve the correct solution to a problem or exercise. In the primary school mathematics curriculum, students meet algorithms as they are taught standard pen-and-paper calculation procedures for addition, subtraction, multiplication and division.

The teaching of standard written algorithms has been a dominant component of the primary school mathematics curriculum for many years. Currently there is considerable debate that questions the value of teaching standard written algorithms. The National Council for Teachers of Mathematics (1989) stated that students should be able to choose between mental, written and technology methods for attaining answers to calculations, with choice dependent upon the purpose of the computation. Being able to choose between methods assumes equal facility and confidence with all three methods. Yet the over-emphasis on algorithms has potentially deskilled students into learning rote procedures with little understanding of why they are doing what they are doing. They have been exposed to teaching practices that prioritise and value the formal algorithm over other methods.

Second, there is now considerable evidence coming from the research on mental computations (Anghileri, 2001; Thompson, 1997) and from the Netherlands (Beishuizen and Anghileri, 1998) which suggests that students should be at liberty to devise their own methods for written computation. The reality is that many students may not be provided with such an opportunity. Many of the calculations undertaken outside the school environment use very different strategies from those used in the classroom. Facility, confidence and competence in these methods would be greatly enhanced if there were a balanced approach to teaching calculations.

The standard algorithms for the four operations are extremely efficient computational procedures that, for many students, have stood the test of time. They are legitimate operations within our base 10 number system and, once learnt, enable computation when the numbers are too large for mental computation and there is no calculator close at hand. Teaching standard algorithms provides children with an efficient means for computation.

Instruction in the algorithms must be carefully planned to enable students to develop an understanding of the steps. Error pattern research has alerted educators to the fact that, for many students, the steps in algorithms have little meaning, and that expert performance is not a measure of understanding (Ashlock, 1994; Resnick, 1982).

In this chapter, the model for teaching algorithms is outlined but we strongly recommend that beginning teachers see this as only one of many approaches to calculating, and suggest that it must be counterbalanced with the use of mental and technological methods.

The value of concrete materials

Concrete materials are typically used to support students' understanding of the steps in the standard algorithm. The use of appropriate concrete materials to embody the steps in the standard algorithms has been the focus of much research. Typical concrete materials for teaching algorithms

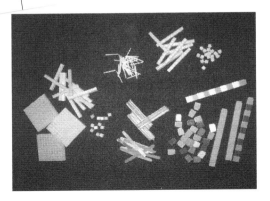

Some of the materials available for teaching algorithms

include sticks that can be bundled into tens with an elastic band, and base 10 blocks. Early models are proportional—that is, size correlates with value—so that the tens block is equivalent in size to ten ones blocks or the ten sticks bundled together represent ten single sticks. These materials are advocated as they can exemplify the processes of the algorithms and assist children to understand the algorithms (Ashlock, 1994; Booker et al., 1980), but their use also places a high cognitive demand on children and can cause confusion through cognitive overload (Bolton-Lewis and Halford, 1992). The how, why and for how long of concrete materials for teaching the algorithms needs to be considered carefully.

Principles for using concrete materials

1. Use concrete materials to support students' understanding of the steps of the algorithm.
2. Carefully select examples for demonstrating the steps in the algorithm—use simple examples that contain small numbers.
3. Once students demonstrate understanding of the steps in the algorithm, as exemplified through the concrete materials, the concrete materials should no longer be used.
4. Use concrete materials to support understanding of the steps only; concrete materials should not be used for the purpose of obtaining solutions.
5. If students require the use of concrete materials to obtain answers, they are not ready to practise written computation methods.

The addition algorithm

● Prerequisite knowledge

The prerequisite knowledge for developing meaning of the standard algorithm is as follows:

1. place value and numeration understanding—that is, understanding of how numbers can be grouped into tens (e.g. 10 and 10 and 10 is 30) and regrouped into tens and ones (25 is 2 tens and 5 ones, or 1 ten and 15 ones) (see Chapter 8)
2. understanding of the concept and language of addition (see Chapter 8)
3. basic addition fact recall (see Chapter 9)
4. estimation, rounding and mental computation skills (see Chapter 9).

● Sequencing

There is a sequence for teaching the standard addition algorithm, as knowledge of adding with particular numbers lays the foundation for adding with other numbers. Learning the written algorithms for adding numbers such as 25 and 32 where there is no trading (or regrouping) between columns lays the foundation for adding numbers such as 36 and 48.

For students who have been encouraged to develop mental methods, teaching the addition algorithm for numbers where there is no carrying may appear to be a useless skill, as calculation of the solution would be a simple mental exercise. However, when using concrete materials to exemplify the written procedure, examples involving small two-digit numbers are required in order to reduce unwieldy manipulation of materials. Once addition with two-digit numbers has been mastered, the procedure can be applied for adding larger numbers, and this is the rationale for teaching the written procedure.

● Linking concrete to written

Materials required for linking the concrete to the written algorithm include place value charts (large enough to accommodate base 10 blocks), base 10 blocks (MAB—see Chapter 8 for a description of this material), pen and paper, and a calculator. As the concrete materials are exemplifying the steps in the written procedure, they must mirror the written procedure as closely as possible. Linking concrete to written for the addition algorithm proceeds as follows:

1. Select a simple example that is easily represented with concrete materials.

Add 36 and 28

2. Write the computation in vertical form, emphasising the importance of aligning the digits in the correct places.

```
   36
 + 28
 ____
```

3. Calculate the answer on the calculator and leave this display visible for ready reference.

4. Display the two numbers to be added (the addends) on the place value chart with concrete materials, arranging the materials to mirror the written representation.

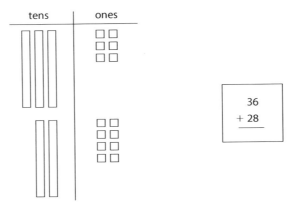

```
   36
 + 28
 ____
```

5. Focus on the ones place in both the concrete and symbolic representation. Verbalise the two numbers and ask students to state the solution, not by counting, but as a basic fact.

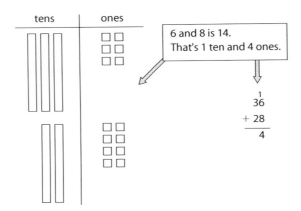

6 and 8 is 14.
That's 1 ten and 4 ones.

```
    1
   36
 + 28
 ____
    4
```

6. If the solution is greater than 10, regroup the ones into tens. Draw attention to how this step is recorded on paper.
7. Repeat steps 5 and 6 as attention moves to the digits in the tens place:
   (a) Verbalise the two numbers, and recall the solution as a basic fact.
   (b) If the solution is greater than 100, regroup the tens into hundreds and tens.
   (c) Draw attention to how this is recorded on paper.
8. Check that the written solution is the same as the calculator display. Repeat this sequence with other examples.

As students manipulate the blocks at the same time as recording in symbols, they are building meaning for the steps in the algorithm. For example, why do you need to start in the ones column when so often we start in the tens column for mental computation? Regrouping provides this reason.

Once students have mastered addition of two numbers, extend this to addition of more than two numbers, but this time, use of concrete materials should not be necessary. The same processes apply as shown in the sequence outlined in this section, which is a further reason for the sequence of steps in the algorithm.

The subtraction algorithm

● Prerequisite knowledge

The prerequisite knowledge for developing meaning of the standard subtraction algorithm is similar to that for addition:

1. place value and numeration understanding—that is, understanding of how numbers can be grouped into tens (e.g. 10 and 10 and 10 is 30) and regrouped into tens and ones (25 is 2 tens and 5 ones, or 1 ten and 15 ones) (see Chapter 8)
2. understanding of the concept and language of subtraction (see Chapter 8)
3. basic subtraction fact recall (see Chapter 9)
4. estimation, rounding and mental computation skills (see Chapter 9).

● Linking concrete to written

In the modelling of the addition algorithm, both numbers to be added are displayed on the place value chart. Modelling the subtraction algorithm is different in that only the minuend (the starting amount) is shown. The subtrahend (the amount taken from the starting amount) is not shown. Materials required for linking the concrete to the written algorithm are the same as for addition: a place value chart (large enough to accommodate base 10 blocks), base 10 blocks (MAB), pen and paper, and a calculator. As the concrete materials are exemplifying the steps in the written procedure, they must mirror the written procedure as closely as possible. Linking concrete to written for the subtraction algorithm proceeds as follows:

1. Select a simple example that is easily represented with concrete materials.

43 take away 28

2. Write the computation in vertical form, emphasising the importance of aligning the digits in the correct places.

```
   43
 − 28
 ____
```

3. Calculate the answer on the calculator and leave this display visible for ready reference.
4. Display only the number from which an amount is going to be subtracted (the subtrahend) on the place value chart.

5. Focus on the ones place in both the concrete and symbolic representation. Are there enough ones to take the required amount

of ones from? If yes, use basic fact recall and record the remaining amount in the ones place. If no, go to the tens place and regroup one ten in that column as ten ones. Place the ten ones in the ones column and subtract the required amount from the ones.

6. Draw attention to how this step is recorded on paper.

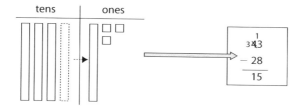

7. Repeat steps 5 and 6 as attention moves to the digits in the tens place.

8. Check that the written solution is the same as the calculator display. Repeat this sequence with other examples.

This process of subtraction is referred to as the *decomposition* method. There is an alternative algorithm for subtraction that many people still remember from their time at school: the *equal additions* method. This method is also sometimes referred to as the 'borrow and pay back' method. This method was taught without recourse to materials for meaning and is not readily modelled with base 10 blocks (MAB). In this method, when a number is too small to subtract an amount from, a '1' is 'borrowed' and written next to the small number in the column under focus, with the '1' paid back to the next number below and to the immediate left. The '1' that was borrowed became a value of 10 and was added to the small number. This then enabled the subtraction to occur. The '1' paid back at the bottom was added (as 1) to the number to be subtracted in the next operation. (For example, 65 – 17; begin with the ones—5 take away 7. Borrow a 1 to make the 5 into 15. Pay back the one to the 7, to become 8). Even though this approach, if followed precisely, ensures a correct result every time, its steps are difficult to model meaningfully. If students in your class know this method and can execute it flawlessly, then there is no need to teach the decomposition method of subtraction as well, as this will cause confusion and is quite a redundant piece of knowledge (unless there is student interest).

● Generalising the addition and subtraction algorithms

The steps in the addition and subtraction algorithms follow a pattern where the digits in the far-right column are dealt with first, moving to the next column to the left of the first digit, until the digits in all columns have been dealt with. An understanding of the legitimacy of the steps within the algorithms promotes conceptual understanding of mathematical principles underlying the algorithms. In the addition algorithm, the ones digits are added to the ones digits, the tens digits are added to the tens digits, and so on.

Similarly in subtraction, the ones digits are dealt with together, as are the tens digits, then the hundreds digits, and so on. It is illegitimate to subtract a ten from a one in the subtraction algorithm; the reason for aligning the digits in their appropriate columns is so that this type of error does not occur. A further principle is that of regrouping across the columns. When there are more than ten ones in the ones column upon adding, groups of ten are 'carried' over to the tens place. This pattern recurs across all places as we proceed from right to left. The principles of the algorithms with whole numbers provide the foundation for understanding algorithms with other types of numbers, such as decimals and fractions; measurements such as time and distance; and algebra, as outlined in the following examples:

1. *Whole numbers*—ones added to ones, tens added to tens, hundreds added to hundreds. Regrouping of ones to tens, tens to hundreds, hundreds to thousands.

    $$\begin{array}{r} 469 \\ + \ 948 \\ \hline \end{array}$$

2. *Decimals*—hundredths added to hundredths, tenths added to tenths, ones added to ones, tens added to tens. Regrouping of hundredths to tenths, tenths to ones, ones to tens.

    $$\begin{array}{r} 25.7 \\ + \ 18.45 \\ \hline \end{array}$$

3. *Fractions*—parts added to parts, wholes added to wholes. Regrouping of parts into wholes.

$$6 \tfrac{1}{4}$$
$$+\ 5 \tfrac{3}{8}$$

4. *Lengths*—centimetres added to centimetres, metres added to metres. Regrouping of centimetres into metres.

```
  59 m 84 cm
+ 26 m 39 cm
```

5. *Time*—minutes added to minutes, hours added to hours. Regrouping of (60) minutes into one hour.

```
  6 hrs 29 mins
+ 4 hrs 35 mins
```

6. *Algebra*—collection of like terms. Like terms are added to like terms.

```
2 + 3y + 7 + 2y
        2 + 3y
      + 7 + 2y
```

The multiplication algorithm

The multiplication algorithm is very complex but extremely efficient. Before students tackle the multiplication algorithm, the following prerequisite knowledge must be in place:

1. place value and numeration knowledge, particularly in relation to understanding the effects upon numbers through multiplying by powers of 10 (see Chapter 8)
2. expanded notation

3. understanding of the concept and language of multiplication (see Chapter 9)
4. basic multiplication facts (see Chapter 9)
5. the distributive property
6. estimation, rounding and mental computation skills (see Chapter 9).

● Place value

The effect of multiplying numbers by powers of 10 provides the foundation for the multiplication algorithm. Entering a single-digit number in a calculator and observing what happens to that number as it is progressively multiplied by ten promotes understanding of movement across the places, rather than a generalisation that zeros are added.

What happens to each number when we multiply by a power of 10?

Th	H	T	O.	t
			6.	
		6	0.	
	6	0	0.	

The number jumps one place to the left and the zero is a place holder.
As we multiply by powers of 10 we move across places to the left.

Using basic fact knowledge and exploring the patterns of the zeros upon multiplication of numbers in multiples of ten promotes number sense in terms of the effects of the operations upon number. Progressive examples enable students to make generalisations about multiplication:

$$6 \times 4 = 24$$
$$6 \times 40 = 240$$
$$6 \times 400 = 2400$$
$$6 \times 4000 = ?$$
$$60 \times 4 = ?$$
$$60 \times 40 = ?$$

As larger numbers are multiplied, the size of the solution becomes anticipated, and this assists in checking the reasonableness of solutions upon using written methods.

● Expanded notation

Expanded notation refers to expressing numbers in their component parts—that is, elaborating the number of ones, tens, hundreds and so on, that are contained in a number. The number 345 in expanded notation is expressed as comprising 3 hundreds, 4 tens and 5 ones (300 + 40 + 5), or $3 \times 100 + 4 \times 10 + 5 \times 1$. The way numbers are represented in symbolic form assumes knowledge of the base 10 numeration system, and that the place in which a digit in a number is located indicates its value. Writing numbers in expanded notation assists in developing understanding of the distributive property.

● The distributive property

In symbolic form, the distributive property is represented as follows: $(a \times b) + (a \times c) = a \times (b + c)$. The meaning of the distributive property can be exemplified using arrays. The following example shows the solution to 3×13 using the distributive property, where 13 is presented in expanded notation as $10 + 3$:

$3 \times 13 = 3 \times (10 + 3)$
$(3 \times 10) + (3 \times 3)$
$30 + 9$
39

3×10 3×3

● Developing the multiplication algorithm

The multiplication algorithm is efficient and concise, and is extremely difficult to teach meaningfully with materials. Sometimes, in an effort to ensure that students understand each step in the process, the algorithm is developed through convoluted and indirect means. If students have facility with all prerequisite knowledge for the multiplication algorithm, presenting them with various models and extended recording methods may result in greater confusion than a more direct approach. A combination of symbolic, concrete and calculator representations to exemplify particular steps in the standard algorithm in a sequential manner is outlined in the following sections.

Simple multiplication

1. Select a simple example that is easily represented with concrete materials.

> 32 multiplied by 3

2. Write the computation in vertical form.

3. Use the calculator (or mental computation) to determine the answer and have this answer displayed for ready reference.

4. Display the number that is to be multiplied on a place value chart.

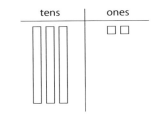

5. Focus on the ones place in both the concrete and symbolic representation. Refer to the written form to check how many times the ones are being multiplied.

6. Multiply the ones by recalling the basic fact.

7. Use base 10 blocks to represent that number.

8. Draw students' attention to the place for recording the numbers of ones.

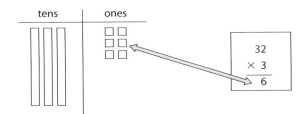

9. Check the ones digit on the calculator display and the number recorded in the ones place.
10. Focus on the number in the tens place in both the concrete and written forms. Refer to the written form to check how many times the tens are multiplied.
11. Multiply the number in the tens place by the multiplier, by recalling the basic fact.
12. Use base 10 blocks to represent that number and draw students' attention to the place for recording the number of tens.

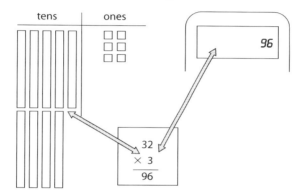

To consolidate the process, repeat with other examples until students can execute the algorithm without the use of the base 10 blocks. Once the pattern has been established, provide examples with larger numbers, such as 124 × 2; 232 × 3; 1124 × 2. When the process of multiplying with single digits without carrying is established, the process of multiplying with single digits involving carrying should be relatively simple, but attention needs to be drawn to the recording of *carried* digits. Continued reference to the calculator display will provide a further check for the legitimacy of the steps in the algorithm. Teaching the multiplication algorithm for numbers involving carrying is described below.

Multiplication with carrying

1. Select a simple example that is easily represented with concrete materials.

| 24 multiplied by 3 |

2. Write the computation in vertical form.

3. Use the calculator (or mental computation) to determine the answer and have this answer displayed for ready reference.
4. Display the number that is to be multiplied on a place value chart.

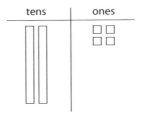

5. Focus on the ones place in both the concrete and symbolic representations. Refer to the written form to check how many times the ones are being multiplied.
6. Multiply the ones, by recalling the basic fact.
7. Use base-10 blocks to represent that number.
8. Draw students' attention to the place for *carrying* the number of tens and recording the numbers of ones.

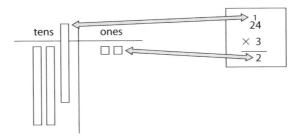

9. Check the ones digit on the calculator display and the number recorded in the ones place. Following the established pattern for multiplying without carrying, ask students to look at the number in the tens place and determine the solution through basic fact recall, emphasising that they must disregard the carried figure. Then add

the carried figure to the basic fact solution and record. Compare the answer with the calculator display.

When multiplying with carrying, the established pattern from multiplying without carrying is drawn upon. For many students, confusion occurs in deciding what to do with the carried digit. During instruction, the teacher must be extremely diligent in ensuring that students fully understand the role of the carried digit, and students should not be provided with practice examples until the teacher is confident of their understanding of that step in the algorithm.

As students become more familiar with the process of carrying in the tens place, provide further examples where the solution requires an answer in the hundreds place (e.g. the solution to multiplying 34×6). Refer to the calculator display to show students how to record the answer.

Multiplying by two-digit numbers

Multiplying by two digits is a case of using prior knowledge of multiplying by single digits and possible new knowledge of multiplying by tens. From experience with number patterns, students should know the effect of multiplying a number by ten. Simple patterning activities, such as those described for place value, should lay the foundation for generalising the effect of multiplying by powers of 10.

The multiplication algorithm occurs in two parts, with the ones multiplied first and then the tens. These two totals are then added. Arrays can assist in providing a visual image of two-digit multiplication.

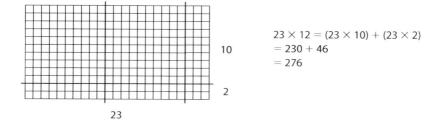

$$23 \times 12 = (23 \times 10) + (23 \times 2)$$
$$= 230 + 46$$
$$= 276$$

When making links between the algorithm and the array, deliberately draw upon students' prior knowledge of the sequence of the algorithm for single-digit multiplication. In two-digit multiplication, the distributive law comes into play. The sequence is to consider the multiplier in extended form (in tens and ones), and then to multiply each digit as in single-digit multiplication.

1. Select a simple example that is easily represented as an array.

2. Write the computation in vertical form.

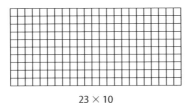

3. Present the computation in array form, exemplifying the distributive property.

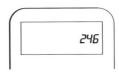

23 × 10   23 × 2

4. Use the calculator to determine the answer and have this answer displayed for ready reference.

246

5. Focus on the digit in the ones place of the multiplier, and on the array in which the product of that number and the multiplicand (the number being multiplied) are presented. Discuss in relation to single-digit multiplication. Complete the multiplication as if for single-digit multiplication and draw students' attention to the recording of the product for that part of the algorithm.

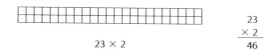

23 × 2

$$\begin{array}{r} 23 \\ \times\ 2 \\ \hline 46 \end{array}$$

6. Focus on the digit in the tens place of the multiplier, and on the array in which the product of that number and the multiplicand is presented. Discuss in relation to multiplication with tens and the fact that the solution will result in a zero in the ones place. Record the zero under the ones place, and complete the multiplication as if for single-digit multiplication. Draw students' attention to the recording of the product for that part of the algorithm.

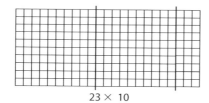

23 × 10

$$\begin{array}{r} 23 \\ \times\ 12 \\ \hline 46 \\ 230 \end{array}$$

7. Consider the array in total and invite students to suggest the next step in the algorithm. Draw students' attention to the recording of the solution. Check the solution on the calculator.

When learning two-digit multiplication, students need to be provided with plenty of time to become familiar with the sequence of steps. The process of the algorithm may need to be revisited several times before students feel confident to execute it unassisted.

The division algorithm

● **Prerequisite knowledge**

The division algorithm is relatively simple compared with the multiplication algorithm. Prerequisite knowledge is similar to that required for the addition and subtraction algorithms, and includes:

1. place value and numeration knowledge, particularly in relation to understanding the effects upon numbers through dividing by powers of 10 (see Chapter 8)

2. understanding of the concept and language of division (see Chapter 8)
3. basic division facts (see Chapter 9)
4. estimation, rounding and mental computation skills (see Chapter 9).

● Place value

The effect of dividing numbers by powers of 10 provides the foundation for the division algorithm. Explorations similar to those for the multiplication algorithm are useful for assisting students to develop understanding of the effects of dividing numbers by powers of 10. Entering a single-digit number in a calculator and observing what happens to that number as it is progressively divided by ten promotes understanding of movement across the places, rather than a generalisation that zeros are deleted (or that the decimal point moves).

Th	H	T	O.	t
	6	0	0.	
		6	0.	
			6.	
			0.	6

The number jumps one place to the right.
As we divide by powers of 10 we move across places to the right.

Using basic fact knowledge and exploring the patterns of the zeros upon division of numbers in multiples of ten promotes number sense in terms of the effects of the operations upon number. Progressive examples enable students to make generalisations about division:

$$24 \div 6 = 4$$
$$240 \div 6 = 40$$
$$2400 \div 6 = 400$$
$$24\,000 \div 6 = ?$$
$$24 \div 4 = 6$$
$$240 \div 40 = ?$$

As larger numbers are divided, the size of the solution becomes anticipated, and this assists in checking the reasonableness of solutions upon using written methods.

● Linking concrete to written

Unlike the addition, subtraction and multiplication algorithms, the division algorithm begins with the digit in the biggest place. The emphasis for the division algorithm is in the use of *sharing* language.

1. Select a simple example that is easily represented with concrete materials.

496 divided by 4

2. Write the computation in standard form.

4) 496

3. Use the calculator to determine the answer and have this answer displayed for ready reference.

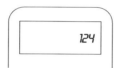

124

4. Represent the number to be divided with base 10 blocks.

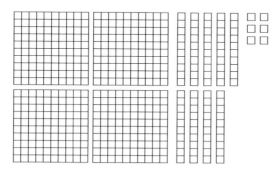

5. Start with the biggest group (400) and calculate the share by recalling the basic fact. Record this solution, drawing attention to the position of the solution and how it indicates the numbers of shares made from that amount.

6. Focus on the next biggest group to be shared, in both the written and concrete forms. Use basic fact recall to determine the number of shares. Draw attention to the position for recording the solution. Emphasise meaning of the steps in the algorithm by drawing attention to the size of the numbers being shared and the placement of the recorded answer at each step. At each step in the algorithm, check the calculator display. Provide further examples until the process is familiar and students feel confident to compute division exercises using the algorithm without reference to the materials. Encourage students to estimate the solution prior to undertaking any computation and to check their resulting answer with their estimated answer. Calculators should always be available for checking results.

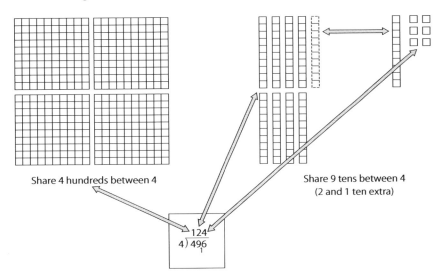

Share 4 hundreds between 4

Share 9 tens between 4
(2 and 1 ten extra)

124
4) 496

Long division

Teaching long division is not a requirement of the primary school syllabus. The long division algorithm has been a topic of considerable debate, and is often cited by students as the turning point at which they began to dislike mathematics. The long division algorithm is simply that—long. It is an algorithm that takes a long time to execute. The long division algorithm can be taught meaningfully, and can promote students' understanding of mathematics and build number sense as well as be an interesting program of study (Lampert, 1992). For many students, however, the long division algorithm may be a piece of knowledge that they do not need. With access to calculators (and many mobile phones have an inbuilt calculator), teaching the long division algorithm should be left to the teacher's professional opinion in accordance with an assessment of the needs of the students.

Teaching written algorithms in the middle years

Teaching children standard written algorithms in the middle years can be a tedious process. The question that must be asked is whether the time and effort required to re-teach standard algorithms is time well spent, or whether promoting number sense and estimation represents a better use of time in the middle years. With calculators readily at hand, this is a serious consideration for the middle years teacher. If instructional time is devoted to teaching written algorithms in the upper middle years, then instruction must be extremely well planned and focused, to ensure maximum impact in minimum time. The examples selected must be well chosen, with MAB used to exemplify the process and not for getting answers. That is, the MAB will be used to legitimise the steps in the algorithm, but students should then practise the algorithm without MAB so that the process can be internalised. However, if students exhibit errors in written algorithms, using concrete materials to demonstrate meaning of the algorithm may not be a simple solution. Research has shown that, despite explicit instruction with appropriate materials, students' errors in written algorithms are extremely difficult to eradicate (Dole, 2003; Resnick, 1982). Other than remedial and intervention instruction, learning new algorithms can be a rewarding problem-solving investigation in the middle years, where students gain further appreciation of the structure and beauty of mathematics. Lampert (1992) has described the interest her Year 5 students showed in learning long division through an investigative approach. The message, then, is to approach the standard written algorithms with caution in the middle years, and question the value for the students in relation to this decision.

Summary

Before teaching algorithms, a substantial amount of prerequisite knowledge must be in place. Recall of the basic facts is necessary to reduce cognitive overload when learning the steps in the algorithms. Estimation and rounding skills are important because they provide a reference point for determining the reasonableness of the calculated solution.

Concrete materials are extremely useful for promoting an understanding of the steps in the algorithms, but they can also be a source of confusion. The use of concrete materials must be carefully considered in instruction for algorithms, and they should never be used by students to obtain answers. Concrete materials exemplify the process, and that is their only purpose.

There is current debate around teaching of standard algorithms. Encouraging students to invent their own algorithms is a trend in mathematics education that is gathering some momentum. Teaching standard algorithms has traditionally occupied a large proportion of the mathematics program. With modern technology, spending this amount of instructional time becomes questionable. However, the standard algorithms are efficient and embody a rich history of development. The task for educators is to consider the impact upon students if standard algorithms continue to be, or are not, explicitly taught.

■ REVIEW QUESTIONS

10.1 Describe the distributive law. Give five examples of computation that are simpler using the distributive law. How does knowledge of this law assist understanding of the multiplication algorithm?

10.2 Provide an argument for teaching the standard addition algorithm in the primary school.

10.3 Outline the prerequisite knowledge required before learning the standard multiplication algorithm.

10.4 What is the purpose of concrete materials in teaching standard written algorithms?

10.5 Imagine you are teaching in a multi-age setting with considerable diversity in achievement levels. Select one of the operations and outline the sequence through which you would teach students.

Further reading

Kamii, C. and Dominick, A. (1997). To teach or not to teach algorithms. *Journal of Mathematical Behavior*, 16(1), 51–61.

Ruthven, K. (1998). The use of mental, written and calculator strategies of numerical computation by upper primary pupils within a 'calculator-aware' number curriculum. *British Educational Research Journal*, 24(1), 21–42.

References

Anghileri, J. (2001). Intuitive approaches, mental strategies and standard algorithms. In J. Anghileri (ed.), *Principles and practices in arithmetic teaching: Innovative approaches in the primary classroom* (pp. 79–94). Buckingham: Open University Press.

Ashlock, R. (1994). *Error patterns in computation: A semi-programmed approach* (6th ed.). New York: Macmillan.

Beishuizen, M. and Anghileri, J. (1998). Which mental strategies in the early number curriculum? A comparison of British ideas and Dutch ideas. *British Educational Research Journal*, 24(5), 519–38.

Bolton-Lewis, G. and Halford, G. (1992). The processing loads of young children's and teachers' representations of place value and implications for teaching. *Mathematics Education Research Journal*, 4(1), 1–23.

Booker, G., Irons, C. and Jones, G. (1980). *Fostering arithmetic in the primary school*. Canberra: Curriculum Development Centre.

Dole, S. (2003). Applying psychological learning theory to helping students overcome learned difficulties in mathematics: An alternative approach to intervention. *School Psychology International*, 24(1), 95–114.

Lampert, M. (1992). Teaching and learning long division for understanding in school. In G. Leinhardt, R. Putnam and R. Hattrup (eds), *Analysis of arithmetic for mathematics teaching* (pp. 221–82). Hillsdale, NJ: Lawrence Erlbaum.

National Council for Teachers of Mathematics (1989). *Curriculum and evaluation standards*. Reston, VA: NCTM.

Resnick, L. (1982). Syntax and semantics in learning to subtract. In T.P. Carpenter, J.M. Moser and T.A. Romberg (eds), *Addition and subtraction: A cognitive perspective*. Hillsdale, NJ: Lawrence Erlbaum.

Thompson, I. (1997). Mental and written algorithms for additions: Can the gap be bridged? In I. Thompson (ed.), *Teaching and learning early number* (pp. 97–109). Buckingham: Open University Press.

CHAPTER 11

Rational number

Topics that fall under the heading of rational number include fractions, decimals, ratio and proportion, rate and per cent. These topics are linked mathematically but conceptually are subtly different. In this chapter, key ideas associated with the topics of fractions, decimals, ratio, rate, proportion and per cent and their interlinked nature are presented. Approaches for enhancing students' knowledge of these topics are described.

Common and decimal fractions

Fractional numbers can be represented in fraction form (e.g. ¼) and in decimal form (e.g. 0.25) and the terms 'common fraction' and 'decimal fraction' respectively are used to distinguish the two symbolic representations. The word *fraction* is frequently applied to numbers in both fraction form and decimal form, yet there are subtle conceptual differences between common fractions and decimal fractions. Common fraction understanding is based on the part-whole concept. Decimal fraction understanding stems from a combination of an understanding of common fractions and whole number and place value knowledge. For simplification, in this chapter common fractions are referred to as fractions and decimal fractions are referred to as decimals.

Whole number and rational number connections

Whole number understanding provides the foundation for understanding of rational numbers. Particular rational number topics provide a foundation as well as a link to other rational number topics. Decimal understanding is connected to both fraction and whole number knowledge. Ratio and proportion understanding links to fractions and also to multiplicative thinking developed through the study of whole numbers. Rate links to ratio. Per cent links to decimals and fractions, and to ratio and proportion. The interconnected nature of rational number topics to each other and to whole number is depicted in the flowchart. The Number Framework (Ministry of Education, New Zealand, 2008) includes nine global stages of number knowledge and strategy understanding that encompass the development of rational number knowledge.

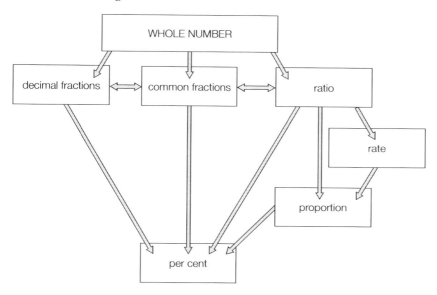

The category of rational numbers includes all numbers that can be expressed in the form a/b, where a and b are whole numbers and b is not zero. From the study of whole numbers, the conceptualisation of numbers falling on a number line that can stretch infinitely to accommodate increasingly larger numbers is promoted. Through the study of fractions and decimals, a further dimension to the number line occurs through locating fractions and decimals between whole numbers on the number line. The link of fractions and decimals to per cent is further consolidated through locating percentages on a number line. Ratio, rate and proportion link to fractions, decimals and percentages, but understanding of these topics is not necessarily enhanced through the number line concept.

Fractions

Student representing one quarter

The basis of fraction understanding is that fractions are a part of a whole—that is, when one whole thing is split into equal parts, each part is a fraction in relation to the original whole. Fractions can also be created from a set of objects numbering more than one. The *whole* in this sense is more than one, but the collective set of objects must be regarded as *one whole* in order to operate with the objects in a fraction sense.

One of the other crucial understandings of fractions relates to fractions as division situations, and this is essential for working with algebraic calculations. Other big fraction ideas are that fractions are ratios, and that fractions are operators. (Fractions as ratio are addressed under 'Ratio and proportion' later in this chapter.) Fractions as operators refers to the way the numerator (the top number) in a fraction is that which multiplies, and the denominator (the bottom number) is that which divides. Knowing fractions as operators should be a natural by-product of a rich conceptual understanding of fractions, and is not elaborated here. In this section, ideas for developing students' understanding of the following three notions are the focus:

- fractions as part of a whole
- fractions as part of a set, and
- fractions as division.

● Intuitive fraction ideas

Many students come to school with intuitive fraction ideas. The most commonly used fraction is one-half, and children are exposed to this from an early age through natural language and experience. When cutting toast or sandwiches, children are often asked if they want their bread cut in half or if they want it left whole. They may be cajoled to eat half their vegetables in order to get ice cream. When they are given a bag of sweets, they are told to give half to their brother. The notion of one-half is usually well established, but mathematically it may not be as precise as required. A young student, provided with a diagram of a rectangle cut into three equal parts, two of which were shaded, was asked to state how much of the rectangle was shaded. The response was 'one-half'. Is it because the student sees two parts of the rectangle are shaded and one part is unshaded that makes it one-half? Or is it because the student names every 'fractional' picture as one-half? Capitalising on students' intuitive mathematics knowledge is important, but ascertaining the mathematical correctness of that knowledge is equally important.

● Beginning fraction activities

Early activities with fractions must be informal and linked to the world of the students. Any situation in which fractions can be identified and the language of fractions used lays the foundation for more formal fraction investigations. When cutting up fruit, the teacher's commentary on the situation would draw students' attention to the fact that the fruit is whole, but it is going to be shared—It is going to be cut up into equal pieces. When it is cut up, there are eight pieces and the teacher emphasises the fact that these eight pieces are part of the whole apple, and they are called eighths. These pieces can be counted (one-eighth, two-eighths, three-eighths . . . eight-eighths—one whole), and counting in eighths shows how the eight pieces can be recombined to make the whole. Through such situations, the students can be exposed to a number of fractional amounts with varying denominators. They can count pieces to complete the whole, and they become familiar with fraction names beyond simply unit fractions (fractions with a numerator of 1).

● Fraction models

To build understanding of fractions, students need to create fractions of various denominators using a range of two- and three-dimensional materials. Examples of various two- and three-dimensional models for the study of fractions include the following:

1. *Region models*—these models are usually regular two- or three-dimensional geometric shapes where all fraction pieces are congruent (the same size and shape).

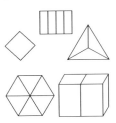

2. *Length models*—examples include string, paper strips, rope, licorice straps. Fractions in this representation lead to locating fractions on the number line and counting in fractional amounts.

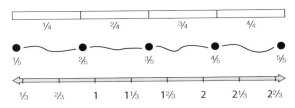

3. *Area models*—these models are similar to region models, but the whole is divided into equal parts that have the same area, though not necessarily the same shape.

4. *Set models*—these include any collection of objects such as counters and blocks. Experience with set models is vital to extend the part-whole concept, and is addressed specifically in the section 'Fractions as part of a set'.

Real-world fraction models

Finding appropriate real-world models for developing mathematical concepts is always an aim in good teaching. However, care must be taken when referring to fractions with real-world objects, for these objects rarely come divided into equal pieces. Is a pizza usually divided into any amount of pieces other than eight (or four for smaller pizzas), and are these pieces ever even? When a birthday cake is divided in the real world, does everyone always get the same amount, or is a smaller amount cut for Aunt Jan because she is on a diet, and a large piece for Uncle Jim because he really likes cake? The typical way in which the real-world item is divided needs to be considered when assessing its value as a model for promoting fraction understanding.

● Fractions as part of a whole

The part-whole concept

Exploration of a variety of fractions with different denominators can be achieved simply with scrap paper, folded and cut or torn to create fractions of various denominators. Such activities provide a platform for discussing whether various representations can show the same fractional amounts.

Half…I'll have the dark half

　　As students create different fractions they are moving through a sequence of five steps:

1. identifying the whole (this is one whole)

2. partitioning into equal parts (5)

3. naming the parts (fifths)

4. selecting the required number of parts (3)

5. naming the fraction (three-fifths).

Each step in this sequence highlights an important aspect necessary for a rich conceptual understanding of fractions. Fractions are parts of a whole (step 1). A fraction is created when the whole is divided into equal parts (step 2). Each part within the whole is named according to the number of parts into which the whole was divided (step 3). The required number of parts is then selected (step 4). The selected parts are named as a fractional amount of the whole (step 5). If three parts out of five are selected, the fraction name is three-fifths. The three indicates the number of parts selected and the five indicates the number of parts into which the whole was divided (see 'Teaching idea—Finding part of a whole').

TEACHING IDEA

Finding part of a whole

- Provide students with sheets of scrap paper (each about half of an A4 sheet of paper).
- Students wave the piece of paper in the air, stating 'this is one whole'.
- Students create various fractions on different pieces of scrap paper (by folding, tearing, cutting, ripping, shading).
- Different representations of each fraction created are checked and discussed in terms of mathematical appropriateness.

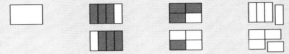

Part–whole scrap paper fractions
- Provide students with sheets of scrap paper (each about half of an A4 sheet).
- Students wave the piece of paper in the air, repeating the fraction name given to the piece of paper by the teacher.

- Students recreate the whole from more scrap paper in terms of the identified fraction part (encourage students to work in pairs).
- Different representations of each whole in terms of initial part are checked and discussed in terms of mathematical appropriateness.

● Finding a whole from a part

Finding a part of a whole from paper can be extended by reversing the activity. The task is for students to recreate the whole in relation to the fraction name given to the piece of paper (see 'Part–whole scrap paper fractions' in the 'Teaching idea'). In order to find the whole in relation to the specified part, students need to consider the number of parts required to make the whole and the number of parts embodied in the piece of paper they hold in their hands. This activity is an invaluable means through which individual students' conceptual understanding of fractions can informally be determined by watching as students interact with each other, ask questions of each other and the teacher, and work with the material.

Fraction family investigations

Manipulation of written fraction symbols has been the dominant approach to formal fraction study, whereby students learnt and practised procedures for fraction computation. Instruction typically involved manipulation of symbols to find common denominators, generate equivalent fractions, reduce fractions to their lowest terms, convert improper fractions to mixed numbers and vice versa. These skills were applied to fraction addition, subtraction, multiplication and division. To many students, the study of fractions is a series of confusing, meaningless exercises, attested to by the range and types of errors students make in fraction calculations (see Ashlock, 1994 for examples of common fraction calculation errors).

Investigations of a family of fractions can assist in providing meaning for written fraction procedures. Four different-coloured paper circles can be used to create a family of fractions, such as halves, quarters, eighths and one whole (see 'Teaching idea—Fraction families'), and manipulated to explore a number of ideas.

T E A C H I N G   I D E A

Fraction families

Halves, quarters, eighths
- Cut out four circles from different-coloured pieces of paper.
- Take one circle and fold and cut it into two equal parts. Write the name of each fraction piece in words on one side and in symbols on the other side.
- Repeat for the second circle, but divide and cut into quarters.
- Repeat for the third circle, but divide and cut into eighths.
- Keep the last circle whole. Write its fraction name in words on one side (one whole) and fraction symbol on the other side (1/1).
- Use the pieces to generate equivalent fractions, to convert mixed numbers to improper fractions and vice versa, to perform simple addition and subtraction calculations, to find common denominators.

Using fraction family materials, explore equivalence, the part-whole concept, improper fractions and mixed numbers, compare fractions and engage in simple addition and subtraction tasks. Students manipulate the materials in response to the teacher's questions. Discussion of the process and students' thinking strategies should be integral to these explorations:

Equivalence
- How many eighths in one-quarter?
- How many eighths in three-quarters?
- One half is equal to how many eighths? How many quarters?

Part-whole concept
- How many quarters make one whole?
- How many eighths in one whole?
- What is one whole less seven-eighths?

Improper fractions and mixed numbers
- How many wholes if I have five-quarters?
- What is six-quarters the same as?
- Eleven-eighths is the same as what?

Comparison
- Which is bigger, ½ or ¼?
- Which is bigger, ⅚ or ½?

Simple addition and subtraction
- How many more pieces to make one whole if I have three-quarters?
- What is three-quarters and three-quarters?
- What different combinations of fraction pieces can you use to make one whole?

Through such explorations, the whole circle remains clearly visible as a reference point. The language of fractions is emphasised through discussion of fraction names, symbolic representation of fraction solutions and symbolic representation of fraction calculations. Encourage students to close their eyes and visualise manipulation of the fraction pieces as various questions are posed to promote mental computation. Different fraction families should only be introduced and explored after students can mentally operate with the first fraction family.

● Fractions as part of a set

For some students, finding one-third of 6, two-fifths of 20, and so on is an extremely difficult task. The difficulty frequently stems from poor fraction understanding and limited work with finding fractions of a whole where the whole is a collection of objects numbering more than 1.

Explorations and investigations using sets of concrete material enhance and extend the part-whole concept to build *part of a set* understanding.

The teaching sequence for developing the part of a set concept is the same as for part of a whole. Using counters or blocks, sets of material in specified amounts are identified as *one whole* (see 'Teaching idea—Developing part of a set concept'). Fractions of that set are found through the process of dividing the set into equal parts, naming each part in relation to the whole, selecting the required number of parts and naming the resulting fractions. Finding part of a set has an added cognitive dimension to finding part of a whole, in that the resulting solution is not a fraction part *per se*, but is a number of individual blocks that can be held in the hand and counted. For example, when ten counters is the whole, two-fifths of ten is four individual blocks that must be seen as two parts of ten. The counters within the set must be disconnected and reconnected to the whole interchangeably.

TEACHING IDEA

Developing part of a set concept

Whole ⟶ part

- Use counters or blocks to make sets of a specified number.
- Identify each set as 'one whole'.
- Find a specified fraction of that set.

Ten counters is one whole   Four counters is two-fifths

Part ⟶ whole

- Use counters or blocks to make sets of a specified number.
- Identify each set as a specific fraction.
- Find a whole in relation to the specified fraction.

Six counters is two-thirds   Nine counters is one whole

To further extend the notion of part of a set, reverse the order of the investigation through finding the whole from a beginning fractional amount. Encourage mental calculation by asking students to visualise the counters in their mind as they explore solutions to given 'part of a set' exercises.

● Fractions as division

Consider the problem '3 divided by 5'. Put it into a 'real' context, and see what happens: share three pizzas between five people. Create three pizzas from paper to act out this situation and discuss how these three 'pizzas' could be shared. Divide each pizza into five pieces and distribute the pieces between three people. The result is that three shared between five gives three-fifths of a pizza each. This adds a further dimension to the conceptualisation of fractions where each symbolic fraction can be regarded as part of a whole, but also as a division. By acting out the situation with the pizzas, attaining the solution shows that the operation is division and that it is partition division (sharing).

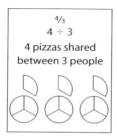

To develop the fractions as division concept, attention needs to be drawn to the fact that a fraction written in symbolic form means the top number divided by the bottom number. This is easier to see when related to improper fractions. The improper fraction ⁴/₃ means 4 divided by 3. The solution is simply 1⅓. The 'rule' for changing an improper fraction into a mixed number takes on meaning through considering the fraction as a different symbolic representation of a division exercise.

As a pizza situation, ⁴/₃ is considered as four pizzas shared between three people where each person gets one full pizza and one-third of the fourth pizza each.

● Fraction computation

Fraction addition and subtraction

Informal addition and subtraction of fractions can be explored through the use of fraction circles, as outlined previously. Use questioning to guide and encourage mental images. Also work with fraction pieces so that students see that there are, for example, four quarters in a whole; five quarters mean that there is one whole and one quarter left over; that one-half is the same as two quarters; and that one-fifth is smaller than one-third.

For simple addition and subtraction, students should be able to *see* the additions and subtractions in their heads, without having to resort to calculation. Pose questions such as the following and ask students about the mental images they see in their mind as they determine the solution:

$$1 - \tfrac{1}{3} =$$
$$\tfrac{3}{5} + \tfrac{1}{5} =$$
$$\tfrac{3}{4} + \tfrac{3}{4} =$$
$$\tfrac{3}{7} + \tfrac{4}{7} = 1 \text{ whole}$$
$$\tfrac{7}{8} - \tfrac{3}{4} =$$

Draw attention to the fraction name and what that means. If adding fifths, how many fifths are required to make one whole? When subtracting quarters from eighths, think of the quarters in an equivalent form with a denominator of eighths.

Fraction multiplication and division

Introducing multiplication and division of fractions needs to be done in a meaningful way that capitalises on prior knowledge. Consider finding $6 \times \tfrac{1}{2}$. This statement makes more sense when linked to the language used for whole number multiplication: '6 groups of $\tfrac{1}{2}$'. One representation of this situation is to skip count in halves along a number line six times to reach a solution of 3. Another representation is to think of halves as half-circles and imagine six of them. Putting two halves together to make one whole, there are three wholes altogether.

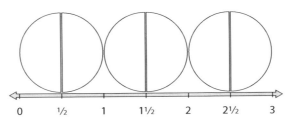

For fraction division (e.g. 6 ÷ ½), referring to prior language of whole number division assists in solution attainment. Rather than state that this is 6 divided by ½, state that the problem is 'how many halves in six'? Ask students to describe or draw the mental picture that comes into their minds as they are asked this question. Do they imagine a number line, where they skip count in halves to 6 to find there are 12 halves? At this stage, simple multiplication and division exercises should be confined to those that can be done mentally.

Decimals

Three notions underpin decimal understanding:

1. Decimals are numbers that can be located on the number line.
2. Decimals are fractions of denominators that are powers of 10 (tenths, hundredths etc.).
3. Place value indicates the value of decimal numbers in symbolic form.

● Decimals as fractions

Prior experience with fractions provides the basis for conceptualising decimals as fractions. The sequence is as follows:

*   exploration of halves, fifths and tenths as a family of fractions
*   symbolic representation of tenths linked to visual image and number line
*   symbolic representation of hundredths and tenths linked to visual image and number line.

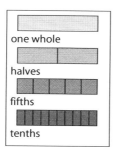

A set of fractions made from paper enables the relationship between halves, fifths and tenths to be explored. When such fractions are created using a linear model, the foundation is laid for visualising decimals on a number line. Encourage manipulation of the fraction pieces until students can determine answers to the following types of questions mentally:

- How many tenths in one whole?
- What is three-tenths and three-fifths?
- How many tenths is the same as one-half?
- What is seven-tenths and nine-tenths?
- What is another name for fifteen-tenths?

Shading pictures of tenths provides a link between the fraction picture, the symbolic fraction form and the symbolic decimal form.

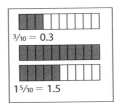

The relationship between tenths and hundreds is seen through creating pictures of various fractions and decimals on 10 × 10 grids. From images created, the relative magnitude of tenths in relation to hundredths is seen. Location of decimals on the number line follows from the pictorial representation.

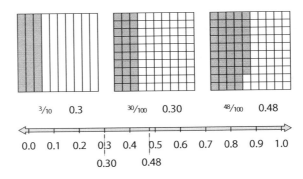

Naming decimals

Many students experience difficulty in reading decimal numbers, in terms of understanding both the size of decimal numbers and the value of the digits within the decimal number. When teachers follow the practice of reading decimal numbers using abbreviated language (e.g. reading 3.4 as 'three point four' rather than 'three and four-tenths'), students are not provided with language support structures in which the meaning of the values of decimal numbers may be emphasised. Pictures of decimal tenths and hundredths provide the link for connecting the symbolic representation to the pictorial representation and also for naming decimals. The equivalence of three-tenths to thirty-hundredths is seen through overlaying a hundreds grid on the picture of three-tenths. The symbolic representation for these two pictures is 0.3 and 0.30 respectively.

In the symbolic representation, the former decimal indicates that there are three-tenths, and this matches the diagram (see above grid). The latter decimal in written form indicates that there are thirty-hundredths, and this also matches the diagram. The decimal 0.48 depicted on the 10×10 grid shows that this number is made up of four-tenths and eight-hundredths, or forty-eight-hundredths. The appropriate name for this decimal is 'forty-eight-hundredths' rather than 'four-tenths and eight-hundredths', and certainly not 'point four eight'. Appropriate naming of decimals during early explorations ensures that the language links to the symbolic representation and the pictorial image.

● Decimals and place value

The symbolic representation of decimal numbers is in accordance with the base 10 numeration system. A decimal number is indicated through the presence of a decimal point. The digit to the immediate right of the decimal point is the amount of tenths in that number; the digit to the immediate right of that digit is the amount of hundredths; and so on. One of the most common misconceptions about decimals is thinking that the more digits to the right of the decimal point, the larger the number (Stacey and Steinle, 1998). For many students, 0.45 is considered to be of lesser value than 0.2476. This misconception stems from poor understanding of decimal numbers and place value.

To understand decimals in relation to place value requires understanding of the following three principles:

1. Places within the base 10 number system extend in an infinite direction to the right of whole numbers to include decimals.
2. The place value system has symmetry around the ones place, but the value of numbers within each place is not symmetrical.
3. The decimal point apparent in the symbolic representation of decimal numbers indicates the ones place and is not a place of value within the place value system.

Through recording decimals in symbolic form and reading decimals for meaning (0.43 is forty-three-hundredths, not point four three), students' understanding of decimals in terms of places in the base 10 system develops. Naming the place signified by each digit in decimal numbers also builds understanding of the symmetrical nature of the places within the number system. The names of the places follow a similar pattern to that for whole numbers, with decimal places taking on the language pattern of fractions (suffix -*th*). The ones place is the centre point for the pattern.

... ten-thousands   thousands   hundreds   tens   ones   tenths   hundredths   thousands   ten-thousandths ...

The relative value of decimals that have many digits after the decimal point is consolidated through practice in locating decimal numbers on a number line. Locating decimals on a number line follows from explorations of fractions of denominator 10 or 100, and counting in decimals. The number 0.5 (five-tenths) is located on a number line halfway between 0 and 1. This requires counting in tenths between 0 and 1. To locate the decimal 0.55 (fifty-five-hundredths) means finding the mid-point between 0.5 (fifty-hundredths) and 0.6 (sixty-hundredths). This requires counting in hundredths between 0 and 1. Extending such an investigation requires rewriting the number line in order for each number to be located, and dividing the space between the numbers 0 and 1 into smaller and smaller sections in groups of 10.

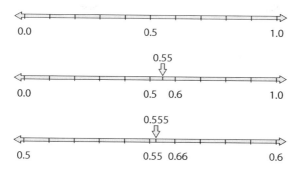

The relationship to 10 between places in our numeration system, established through experiences with whole number study, needs to be reinforced when decimals are introduced. Exploring the effect upon numbers through progressively multiplying and dividing by 10 assists in consolidating this notion. Entering a number on a calculator and observing what occurs to the calculator display of that number as it is progressively multiplied by 10 shows how digits move across places. Keeping a record of this exploration provides a reference for discussion, particularly when the numbers are recorded in an organised list under place value columns. The number 43, when multiplied by 10, becomes 430. When it is multiplied by 10 again, it becomes 4300. Each digit in the number moves one place to the left and its value is ten times greater than its original value. Upon division by 10, the digits retreat to the right and eventually the decimal point appears to separate the ones from the tenths. The value of each digit upon successive multiplication or division is 10 times more or 10 times less than its previous value. Through this investigation the display of zeros on the calculator can be discussed. A calculator does not show excess zeros beyond the decimal point.

	Th	H	T	O.	t	h	th
			4	3.			
× 10 ⟶			4	3	0.			
× 10 ⟶		4	3	0	0.			
÷ 10 ⟶			4	3	0.			
÷ 10 ⟶				4	3.			
÷ 10 ⟶					4.	3		
÷ 10 ⟶					0.	4	3	
÷ 10 ⟶					0.	0	4	3

Exploring the pattern that occurs when dividing numbers by 10 highlights the place and value of decimal numbers within our numeration system. It emphasises how each place in the numeration system is linked by powers of 10 and that each place to the right is ten times bigger than the place before. Similarly, each place to the right is ten times smaller than the place before. When interpreting decimal numbers, tenths are ten times greater than hundredths, and hundredths are ten times greater than thousandths, and so on.

Observation of the movement of digits across the places upon multiplication and division by 10 provides students with an alternative to the commonly held belief that the decimal point moves. It is the *digits* that move—to the left upon multiplication by 10 or powers of 10 and to the right upon division by 10 or powers of 10.

Precision and error in rounding decimals

Through experience with decimals, students come to realise that the more digits there are after the decimal point, the more precise the number, but also the increasing insignificance of the value of these digits. When calculating with decimals, it is common for students to be instructed to round their answers to the nearest tenth or hundredth. Knowing the relative value of digits after the decimal point gives meaning to rounding, and also places students in a position to decide the number of places to which a decimal number should be rounded. In the real world, inappropriate rounding of decimals to tenths or hundredths may be detrimental. In medical experiments and operations, inaccurate calculating and measuring may cause risk to patients. In the building industry, machines incorrectly calibrated could result in damage to equipment. In the school classroom, problem-solving and calculations with decimals should be contextualised so that rounding of decimals is appropriate to the situation.

Ratio and proportion

The topics of ratio and proportion are commonly discussed simultaneously, and indeed are conceptually linked. In its simplest form, *ratio* describes a situation in comparative terms; *proportion* is when this comparison is used to describe a related situation in the same comparative terms. For example, if we say that the ratio of boys to girls in the class is two to three, we are comparing the number of boys to the number of girls. When we know that there are 30 children in the class we know that, proportionally, the number of boys is twelve and the number of girls is eighteen. We are using the base comparison to apply it to the whole situation. In order to understand this relationship, proportional reasoning is used. Proportional thinking and reasoning involve knowing the multiplicative relationship between the base ratio and the proportional situation to which it is applied.

Like all mathematics topics, developing proportional reasoning and thinking skills takes time and is dependent upon sound foundations of associated topics, particularly multiplication, and having rich conceptual understanding of ratio and proportion. Ratio and proportion are two concepts that present a challenge to both teachers and students. Proportional reasoning is fundamental to understanding many topics in the middle school curriculum, such as the geometry of plane shapes, trigonometry and particularly percentages, as well as rate, ratio and proportion applications. As Lesh and colleagues (Lesh et al., 1988) state: 'Proportional reasoning is the capstone of children's elementary school arithmetic and the cornerstone of all that is to follow.' (pp. 93–4).

Traditionally, developing ratio and proportion was considered a topic for upper primary and secondary grades. However, teachers at all year levels need to provide students with rich experiences and appropriate activities in order for them to develop understanding of these topics, so that later they will be able to apply this knowledge in a meaningful way. Teachers of early primary classes must ensure that the seeds of the concepts are sown in their students' minds in the early years, and as students move through the grades, all teachers must build on and extend this knowledge.

● Ratio and fractions

Ratios are commonly described as fractions. In the true mathematical sense, fractions are ratios in that they are rational numbers that can be expressed in the form a/b (provided that b does not equal zero). However, the general meaning of fraction as being part of a whole (as described in the previous section on fractions and decimals) interferes with the conceptual meaning of ratio. Continuing with the boys and girls example above, it was stated that the ratio of boys to girls is two to three—that is, for every two boys there are three girls. When this is considered in the context of the whole class, there is a base unit of five children, of which two are boys and three are girls. The boys and girls situation describes the two parts of the whole unit of children. In this sense, the ratio of boys to girls describes a part-to-part relationship, whereas fractions typically represent a part-to-whole situation. In a fraction sense, the class is made up of two-fifths boys and three-fifths girls.

T E A C H I N G   I D E A

Ratio language through patterns

Ratio language through patterns

■■□■■□■■□
For every 2 black there is 1 yellow

●●●■●●●■
For every 3 circles there is 1 square

Ratio describes a comparative relationship between two entities. Ratio can be used to describe part-to-part, whole-to-whole and part-to-whole situations:

*   part-to-part ratio (e.g. comparing the number of girls in a class to the number of boys in the class)
*   whole-to-whole ratio (e.g. comparing the number of girls in one class to the number of girls in another class)
*   part-to-whole ratio (e.g. comparing the number of girls in the class to the number of children in the class).

Predominantly, ratio is used to describe a part-to-part relationship. The misleading nature of expressing the ratio of boys to girls in fraction form (in the illustration above as ⅔) highlights the importance of considering when the symbolic representation for ratio situations should most appropriately be introduced to students.

● Starting points for ratio and proportion understanding

Through experience and play, learners encounter ratio at an early age. As they get dressed in the morning and are searching for their socks, they know that for every one sock there is another sock; that for every shoe there is another shoe. When they help set the table for dinner, they know that for each person there is a fork and knife; for every soup bowl there is a spoon. When playing 'tea parties' with toys, there is one cup and one saucer for each doll. When constructing patterns with blocks, there is one red block for every two blue blocks; or two greens for every yellow. The link between multiplication and division is evident in these examples: if there are five dolls, there are ten cups and saucers; if there are six knives and forks, this is sufficient for three people. Early ratio investigations provide an opportunity for immersing learners in the language of ratio. If it takes three shovels of sand to fill the truck, it will take six shovels of sand to fill two trucks.

Copying, creating and extending patterns with blocks or other material provides a natural environment for ratio language and for posing questions that promote multiplicative thinking:

- What blocks do we need to continue the pattern?
- How many blue blocks for every yellow block?
- How many blue blocks would we need to extend the pattern if we had three yellow blocks?
- How many blue and how many yellow blocks would we need if we had twelve blocks to build the pattern?

Such questions provide a challenge to young minds to consider the base unit of the pattern. Posters of patterns that indicate the base unit of the pattern provide a permanent record of ratio situations.

Guided ratio investigations in the early years can occur through making mixtures. Mixtures that involve edible or drinkable ingredients (such as cake mixtures or fruit juice punch) can be used to appeal to students' sense of taste, and how increasing or decreasing particular elements within various mixtures will alter the taste accordingly. The activity presented in the 'Teaching idea—Making punch mixes' provides opportunities for the promotion of proportional reasoning, through such questions as:

- If perfect punch is made from two cups of orange juice and one cup of lime juice, how many cups of orange juice would be needed to make the ideal punch mix if you had eight cups of lime juice in the mixing container?
- What if it was discovered that the ideal mix for punch was to use one cup of lime juice for every three cups of orange juice? How many cups of each juice would be required to make punch for a party so that the twelve children each get one cup of punch?

These questions focus on different dimensions of ratio understanding:

- When new mixes are made from a base ratio, they taste the same.
- In any mixture, increasing the number of cups of one mixer requires that the second mixer must also be increased, in accordance with the base ratio, so that the mix tastes the same.
- The two mixers are two parts of the ratio. Adding the parts gives the total number of cups of the base mixture.

TEACHING IDEA

Making punch mixes

Do these two mixes taste the same?

Mix 1:

1 cup lime juice

1 cup orange juice

Mix 2:

2 cups lime juice

2 cups orange juice

Make up the two mixes and pour into their own bowl. Taste them.

1. Do the mixes taste the same?
2. How many cups of lime juice would be needed if there were 4 cups of orange juice?

3. How many cups of each mix is required to mix enough punch for 12 children?

Make up some of your own mixes.

Draw pictures of your mixes.

Make a mix for a party of 24 children. How much of each mixer do you need?

In a mathematical sense, the different dimensions of ratio understanding promoted through mixing activities relate to:

- Ratio equivalence (e.g. 1:3 = 2:6)
- Part-to-part and the multiplicative structure of ratio.
- Part-to-part and the relationship to the whole.

In early ratio investigations, the teacher's role is to engage students in discussion, problem-solving and reasoning in a guided sense. Through such explorations, the students can explore the sophisticated topic of ratio and developing proportional reasoning skills through discussion, hypothesising and justification without using mathematical symbolism.

Promoting language and mental images of ratio and proportion

As students engage in hands-on ratio explorations, a natural context for the use of comparative language to describe ratio situations is provided. The word 'part' is used to label the components in the ratio, and the phrase 'for every . . . there is . . . ' is used to describe the relationship between the parts in the ratio. When making mixes, the mix is made up partly of one type of mixer and partly of another type of mixer: 'The mix is part orange juice and part lime juice. For every two cups of orange juice there is one cup of lime juice. For every two parts of orange juice there is one part of lime juice.' Through the use of such language, the word ratio can be introduced in a meaningful sense, and also as a shorthand way of describing the situation: 'The mix of orange juice to lime is in the ratio of 2 to 1.' In early ratio explorations, the symbol for ratio (:) should be avoided at all costs, but if it is used it should be accompanied with the names of the items being compared (the ratio of orange juice to lime juice is 2:1).

● Visual images and mental computation

As students become comfortable with the language of ratio, they need to be provided with activities that promote the development of mental images that assist in mentally carrying out simple ratio and proportion calculations. The three types of ratio calculation students should be able to perform mentally are:

1. generating proportions from simple ratios using multiplication
2. finding simple ratios from proportional situations using division
3. determining missing elements of ratio and proportion situations through part-part-whole concept knowledge.

The following activities provide a sequential progression for assisting mental computational proficiency. The activities utilise coloured counters displayed on ratio tables to promote strong visual images of the process of the calculation.

Generating proportions from simple ratios using multiplication

Simple ratio calculation can be performed when a unit ratio is given (one of the numbers in the ratio is 1). The unit ratio is the basis for generating equivalent ratios, or proportional amounts. This can be exemplified with concrete materials. Using counters of two different colours (e.g. red and blue), one red counter and two blue counters show a ratio of 1 to 2; that is, for every one red counter, there are two blue counters. Proportions can be generated by adding more counters in the pattern of the original ratio—that for every red counter there are two blue counters. When there are four red counters, there must be eight blue counters. Placing counters in a ratio table, consisting of columns headed with the comparative parts in the ratio, assists in organising the counters and exemplifying the multiplicative nature of proportions. Other ratios can be explored through teacher questions focusing attention on students' prior knowledge of multiplication to generate proportions, for example:

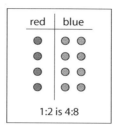

- If red and blue counters are in the ratio of 1:4, how many blue counters will there be if there are 3 red counters?
- If red and blue counters are in the ratio of 5:1, how many red counters will there be if there are three blue counters?

red	blue
● ●	● ● ●
● ●	● ● ●
● ●	● ● ●
● ●	● ● ●

2:3 is 6:9

To encourage mental computation, ask students to close their eyes and visualise counters on a ratio table as they calculate proportional amounts from unit ratios. Engage students in discussion about their mental images and strategies, and encourage them to describe their thinking in terms of multiplication. A simple checking exercise will ascertain students' level of comfort with the notion of generating proportions from unit ratios: ask them to mentally complete missing numbers of given proportions (e.g. '1 to 4 is the same as 2 to ?'; '2 to 1 is the same as ? to 10'; '1 to 3 is the same as ? to 12'; and so on).

Exploring ratios that are not unit ratios (where neither of the numbers in the ratio pair has a value of 1) is an extension of exploring unit ratios. As in the preceding activity, counters on a ratio table provide the visual image of the ratio and assist in generation of the proportion (e.g. for the ratio of 2 to 3, there must be nine blue counters when there are six red counters because we multiplied 2 by 3 to give six red counters, so we must multiply 3 by 3 to get nine blue counters).

Simplest form ratio through division

Proficiency with mentally generating proportions from unit ratios lays the foundation for finding simplest form ratios; this is actually the reverse of the process of generating proportions from simple ratios. To orientate students to this notion, build on their skill of visualising counters on the ratio table, such as red and blue counters in the ratio of 3:9. Use counters to represent this situation, encouraging students to place the red counters in a single vertical line and to distribute the blue counters evenly between the red. The division process is reinforced through this action. The counters clearly show that for every red counter there are three blue counters. From a proportion of 3 to 9, the simplest form ratio is 1 to 3.

The exploration of many different ratios with concrete materials assists in building students' confidence in the process and familiarity with the language, and should be continued until students can readily calculate simplest form ratio with diminishing reference to the counters. Encourage students to verbalise the situation as a ratio before they separate one group of counters into a vertical line and then distribute the other counters equally between the separated counters. Draw upon students' number sense before counters are distributed by considering common factors between the two numbers in the ratio. Thinking of ratios in terms of common factors links to the skill of finding simplest form fractions and consistently dealing with all parts in the situation. Encourage mental computation through asking students to close their eyes and visualise the counters on the ratio table as they determine the simplest form ratio.

4:6

common factor of 2, divide each number in the ratio by 2

red	blue
● ●	● ● ●
● ●	● ● ●

4:6 is 2:3

red	blue
●	● ● ●
●	● ● ●
●	● ● ●
●	● ● ●

3:9 is 1:3

Promoting the part-part-whole structure of ratio situations

In the previous ratio explorations, the focus was on generating equivalent ratios (proportional amounts) or describing a collection of objects as a ratio in its simplest form. This lays the foundation for explicitly promoting the part-part-whole concept of ratio situations and consolidating the language of ratio.

Place red and blue counters on a ratio table to show a ratio of 2:3 (for every two red counters there must be three blue counters). Focus students' attention on the part-part-whole nature of the situation through direct questioning:

- How many red?
- How many blue?
- How many counters in total?
- If there were a total of ten counters in the ratio of 2:3, how many of each colour would there be? Four red and six blue.
- Why? Ratio 2:3 means five counters in total. If ten counters are used, I need to double each part.

Explore the part-part-whole nature of other ratios by working through the same sequence of questions (e.g. in a ratio of 2:3, how many counters of each colour are there if there are fifteen counters?; in a ratio of 5:1, how many of each colour are there if there are eighteen counters?). Promote mental computation by encouraging students to determine solutions by visualising the situation.

● Problem-solving and ratio calculations

It often happens in ratio instruction that students are provided with problem situations that can be solved easily through multiplication or division. Take the example of finding the amount of fertiliser to be mixed with 15 litres of water in the ratio of two parts fertiliser to three parts water. From the base ratio of three parts water and comparing it to the 15 litres of water in the given amount, it can be seen that the 3 has been multiplied by 5 to give 15, so the solution is easily acquired by multiplying the 2 by 5 to give 10. When given ratios and amounts are not simply related, the previous strategy does not provide many clues for action to solve particular problems. Consider the example of finding the amount of fertiliser that would be required if 13 litres of water were used. The calculation methods offered to students must lay the foundation for all types of situations that they might encounter, and these methods must be presented in a meaningful way.

Encouraging students to use ratio tables for all ratio explorations provides a solid foundation for promoting the multiplicative thinking necessary for solving 'ugly' ratio problems. As shown in the examples on the next page, the *easy* ratios are a process of multiplication and recognition of number patterns and relationships. In dealing with 'ugly' ratios, the process of converting the given ratio into a unit ratio is tracked through the ratio table.

Easy ratio   'Ugly' ratio

Ratio tables in the vertical sense are shown above. Ratio tables can also be constructed in a horizontal sense. Horizontal ratio tables have been proposed as a means for assisting students to develop mental strategies for solving proportion problems (Middleton & van den Heuvel-Panhuizen, 1995). Ratio tables encourage the use of number strategies such as halving, doubling, multiplying by 10, and so on. Progressive and simultaneous operation on the given numbers shows how the relationship (ratio) is preserved proportionately. The following example shows how a ratio table can be used to calculate how many seedling plants are packaged into 14 boxes, when each box holds twelve plants.

- *Step 1:* Construct the ratio table and display the given information.

Boxes	1				
Seedlings	12				

- *Step 2:* Select successive operations (e.g. multiplying by 10, halving, doubling) to determine the solution. Use arrows above each number to show the journey to the solution.

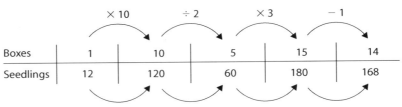

Boxes	1	10	5	15	14
Seedlings	12	120	60	180	168

In this example, the pathway to the solution was to multiply by 10, then divide by 2, then multiply by 3 to get to 15. This is one more than the required number, so then just subtract 1 box of seedlings (12) to reach the solution of 168.

In the next example, drinks are packaged into fifteen bottle boxes. How many drinks would there be in sixteen boxes?

- *Step 1:* Construct the ratio table and display the given information.

Boxes	1				
Seedlings	15				

- *Step 2:* Select successive operations (e.g., multiplying by 10, halving, doubling) to determine the solution. Use arrows above each number to show the journey to the solution.

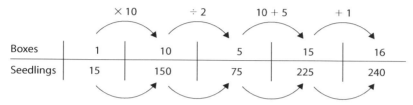

In this example, the pathway to the solution was to multiply by 10, then divide by 2, then add 10 and 5 to get to 15. This is one less than the required number, so then just add one box of drinks (15) to reach the solution of 240.

These two examples show different pathways that can be taken to arrive at the solution. In the first example, after multiplying by 10 and then halving, the next step was to multiply by 3. In the second example, after the amount for 10 and 5 had been determined, these two amounts were added. This required looking back at the table to locate previous calculations that assisted in reaching the destination.

To use ratio tables effectively, students need lots of practice. They also need support and guidance to ensure that the ratio table is used as an effective tool for ratio calculations. Students need to be reassured that they can extend the ratio table as far as required to arrive at a solution. They also need to be reassured that they don't need to fill in all the cells if they have reached a solution and there are empty cells. They also need to overcome their perception that ratio tables must be constructed 'in order'—that is, that after multiplying by 10, they can readily halve the amount in the next cell and it is not 'out of order' in the sequence.

Rate

The topic of rate is usually associated with ratio. Many of the procedures associated with ratio calculations also apply to rate situations. Rate

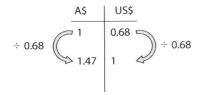

A$–US$ exchange rate mid-2003

describes a relationship between two different measures: cents per litre, dollars per kilogram and metres per second are examples of rates.

Just as ratio tables assist in organising information for ratio investigations, they can also be used to assist in calculation of rates. The use of ratio tables provides a structure for organising the components of the comparative situation and for tracking operations on the numbers during the solution process. With such a structure, problem-solving with rates and ratios becomes a manageable task, and even currency exchange between countries becomes simple to visualise, conceptualise and calculate. For example, a ratio table can assist in conceptualising the exchange rate of US dollars to Australian dollars. As seen in the ratio table depicted, at the mid-2003 rate A$1.00 was equivalent to US$0.68. To make the US dollar the focus of investigation, divide 0.68 by 0.68 to give 1. Correspondingly, the value in the Australian dollar (A$) column on the ratio table must then be multiplied by 0.68. This gives a total of $1.47. Thus the US$1, when converted to Australian dollars at the current rate of 0.68, is equivalent to A$1.47.

In 2001–02, the Australian dollar compared to the US dollar was approximately 0.56. That is, A$1.00 could purchase 56 cents in the United States. Hence, for every US dollar, you would receive $1.78 in Australian currency at this rate. In 2010, the Australian dollar reached parity with the American dollar. This was great news for Australians travelling to and shopping in the United States. In Australia, in 2010, Converse sneakers were A$90. In the US, they were selling for US$60. If the US dollar was at the 2001 rate, the Converse sneakers would be A$107.14 and hence not much cheaper to buy. However, in 2010, the US price meant a great saving to Australians. You can track these calculations by constructing a ratio table in a similar manner to the previous example. For example, if a car travelled 60 kilometres in half an hour, its speed could be calculated as distance (60 km) divided by time (0.5 hour) to give a speed of 120 km/hour. The best-known rate is speed. Often the definition of speed is given as distance over time and the process of division is used to calculate speed given measures of distance and time. Thinking of rate as merely a division process limits its

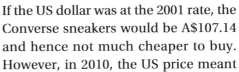

Time (hr)	0.5	1	1.5
Distance travelled (km)	60	120	180

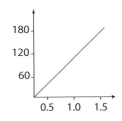

Graph showing speed of a car that travels 60 km in 0.5 hours

conceptualisation as a linear relationship between two measures—this can be more appropriately conceptualised in graphical form.

Rate is used frequently in our society so investigations using rate can easily be linked to the real world. Common rates are found in the purchase of goods in dollars per kilogram, petrol in cents per litre, mobile phone calls in cents per minute (or second), part-time employment in dollars per hour. Showing rates as a graph reinforces the relationship between the two measures in the rate.

Per cent

Evidence of per cent usage and applications abounds in our society. In shopping we are confronted with discounts; in banking we are presented with interest rates; and in media reports per cent is used to convey information. However, a simplistic view of per cent is often the reason for its misapplication, as the following common errors indicate:

- A newspaper article reported the story of a statue picked up for $1 at a garage sale and then being valued at $200, with the headline reading that the lucky statue owner had made a 200 per cent profit.
- Train fares increased 300 per cent, from $3 to $9.
- To encourage sales but also to insure against loss, the store owner increased prices by 25 per cent before advertising a discount of 25 per cent on the new price.

Per cent is a difficult topic to teach and to learn, but is often taught in a narrow fashion to only link to fractions and decimals (Parker and Leinhardt, 1995). A rich conceptual understanding of per cent and confidence to use and apply per cent relies on knowledge of not only fractions and decimals, but also ratio understanding and proportional reasoning skills. Students who can successfully operate in the domain of per cent are those with well-developed 'per cent sense' (Dole et al, 1997).

● Meanings of per cent

Per cent has many meanings and is used for different purposes in different contexts. The multifaceted nature of per cent is seen in the following meanings:

- Per cent is a fraction: 35% is $^{35}/_{100}$ in the part-whole sense.
- Per cent is a decimal: 35% is 0.35.
- Per cent is a number: 35% can be counted and located on a number line.
- Per cent is a ratio: there are 35% more girls in Class A than in Class B.
- Per cent is a proportion: 70 out of 200 is equivalent to 35 out of 100, which is 35%.

Per cent is a difficult concept because of its many meanings, and the many meanings are masked by the single concise symbol: %. Per cent instruction needs to promote students' understanding of the various applications of per cent; this is promoted through explicit teaching of per cent as a ratio and proportion, and through building upon and linking to students' prior knowledge of fractions and decimals (Dole, 1999, 2000).

● Building conceptual understanding of per cent

The basic concept of per cent is its link to a base of 100. Linking per cent to fractions and decimals is an extension of using 10×10 grids to represent decimal (and fraction) numbers (as outlined in the section on decimals). As children draw various pictures of fractions with denominators of 100, labelling the diagram as a per cent shows the equivalence of fractions, decimals and per cent. Drawing particular percentages in a highly visual way can also show the link of some common percentages to fractions (e.g. 50% is ½, 25% is ¼, 75% is ¾, 10% is ¹/₁₀) and promote per cent benchmarks (see the following diagram).

25% as 2 × 10% + 5%   25% as ¼   20% as ⅕

The difficulty with using 10×10 grids for per cent representation arises when percentages are greater than 1. To draw a picture of 115 per cent, the dilemma is whether to use two grids when one is not enough, which can possibly evoke the misconception of two wholes being required for percentages greater than 1. A better way is to use

number lines simultaneously with 10 ×
10 grids. For percentages greater than 1,
the number line can be extended beyond
the 100 per cent mark. Representing
percentages on a number line is a means
of promoting understanding of per cent
as a number.

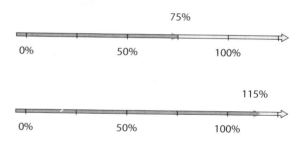

● The language of per cent

Per cent is frequently used in everyday language and conversation,
but quite often its usage does not adhere to the mathematical rules
of per cent. This may be a source of confusion for students. It is not
uncommon for young children playing sport to be told by the coach
that they are expected to put in '110 per cent effort' for the game. If
100 per cent is used to designate the whole, can the whole be more than
100 per cent? Engaging students in discussion of the language of per
cent in colloquial terms and mathematical terms is a valuable way of
determining their intuitive notions of per cent and the extent to which
these notions may help or hinder further conceptual development of
per cent in a mathematical sense.

TEACHING IDEA

The language of per cent*

In groups, discuss what the following phrases mean and when they
might be used.

Can we have:
   50% class attendance?
   90% class attendance?
   100% class attendance?
   110% class attendance?

Can you be:
   50% certain?
   100% certain?
   110% certain?

How much effort is:
   50% effort?
   10% effort?
   100% effort?
   110% effort?

What do the following mean?
   75% full
   25% full
   100% full
   25% empty
*Adapted from Glatzer (1984)

● Per cent part and complement

To promote a rich understanding of per cent, students need to think of per cent as describing a situation in terms of a base of 100. They also need to see the situation as one part in relation to 100 and the remaining part (the complement) in relation to 100. For example, if a label on a wool sweater states that the sweater is 80 per cent wool, then it is known that 20 per cent of the sweater is made of fibres that are not wool; if 55 per cent of the people voted for no homework, then 45 per cent of the people voted for homework.

To assist students to visualise per cent situations in terms of the part and the complement, it is useful to represent various situations on 10×10 grids. By shading the various parts of the whole on the grid, a visual image is presented, and the part and complement are seen as comprising the whole. Shading 10×10 grids is simpler than using a circle graph for representing information in percentages, as the 100 parts in the whole are clearly visible.

Encourage mental computation of simple per cent situations and practice of calculations to 100 by asking students to close their eyes and visualise various per cent part and complement situations. Prior experience of depicting per cent situations on 10×10 grids provides a strong visual image for assisting in determining solutions to such things as:

- If I drank 25% of the milk, what percentage was left?
- 28% of the smarties were green; what percentage was not green?
- The jacket was discounted by 25%; what percentage of the original price did I have to pay?
- Rump steak is 82% beef; what percentage is fat?
- Soap contains 0.5% perfume, 37% pure soap and the rest is fat. What percentage is fat?
- The human body is about 78% water. What percentage is made up of other stuff?

● Mental computation of per cent situations

The development of common per cent benchmarks (e.g. 50% is one-half; 25% is one-quarter; 10% is one-tenth) lays the foundation for simple mental computation of finding a percentage of an amount. Per cent/ fraction equivalence is useful for supporting mental computation. For example, to find 25% of 80, it is simple to think of 25 per cent as one-quarter and then to divide 80 by 4. Similarly, thinking of 10 per cent as one-tenth assists in mentally dividing the amount by 10 in order to find 10 per cent. Finding a percentage of an amount is the simplest of all per cent calculations, and students need to be provided with opportunities for mentally calculating simple percentages of this type. Use selected examples to support student facility with calculations of percentages: 25% of 800—think one-quarter of 800; 25% of 650, think half of 650, then half again.

● Increases of more than 100 per cent

The three misapplications of per cent presented at the start of this section stem from misunderstanding of per cent increases of more than 100 per cent. The difficulty in conceptualising and checking calculations of increases of more than 100 per cent lies in the potential confusion caused by the fact that the situation can be considered from both an additive and a multiplicative perspective. The first example given at the beginning of this section describes incorrect reporting in a newspaper article of a statue purchased for $1 and then being valued at $200 and stating that this was a 200 per cent increase. The confusion is that the original whole amount ($1) has not been considered as 100 per cent, but has possibly been considered as a value of 1. As the original amount of $1 is equivalent to 100 per cent, an increase of 100 per cent would mean doubling the original amount to give a new whole of $2. A 200 per cent increase is actually three times the original amount, which in this case would be $3. By progressively considering percentage increases in this way, the magnitude of the actual percentage increase to reach $200 becomes apparent. The actual percentage increase from $1 to $200 is 19 900 per cent.

TEACHING   IDEA

Exploring increases of 100 per cent

Take a group of counters.
Encase in the hands and state: 'This is one whole.'
Increase the group by 100%.
State how many counters in the new group.
State this additively as a per cent increase.
State this multiplicatively as the number of times bigger the new group is compared to the original.

Increase by 200%

4 counters

A 200% increase means 3 times the original

Considering 100 per cent and 200 per cent increases as in the example above, the additive and multiplicative interpretations of per cent increases can be seen. If the statue had been purchased for $1 and resold for $2, the per cent increase from an additive interpretation would have been 100 per cent. That is, 100 per cent has been added to the original whole. In a multiplicative sense, the new price is 200 per cent of its original price. In the train fare example given at the start of this section, an increase from $3 to $9 means that the fare has been increased additively 200 per cent, or multiplicatively as 300 per cent, from the original price. The subtle differences between interpreting increase situations in a multiplicative or additive sense help to explain how simple errors can be made.

The multiplicative and additive interpretations of percentage increases of 100 per cent and other multiples of 100 can be assisted through explorations with concrete materials. Take a small group of counters and increase the number by 100 per cent, by adding the same amount of counters as the original group. The size of this group is then compared to the original amount (see 'Teaching idea—Exploring increases of 100 per cent'). An increase of 100 per cent means the new size of the group is twice that of the original amount. When the group is increased 200 per cent, the increase is three times the original. Through such explorations, the teacher's role is to encourage students to generalise what happens as a collection is increased by an amount of more than 100 per cent or multiples of 100 per cent. Draw students' attention to the fact that, for example, a 200 per cent increase is not the same as multiplying by 2. Such investigations lay the foundation for visualising mental calculations of amounts greater than 100 per cent. Ask students to close their eyes and visualise increasing groups by various percentages (e.g. 4 counters, increase 100%; 50 counters, increase 100%; 800 counters, increase 300%). Always encourage students to discuss their solution strategies.

● Per cent increases and decreases

For exploring per cent increases of more than 100 per cent, the language foundation and multiplicative structure is emphasised in order to use the language of per cent to describe situations appropriately. Refer to the example given at the start of this section of the shop owner increasing prices before a sale by 25 per cent and then giving a 25 per cent discount. Does the shop owner win, lose or break even in this situation? In fact, by employing this strategy the shop owner loses. Modelling this situation with counters exemplifies why this is so. Take a set of eight counters. Increase the group by 25 per cent (add two counters). The new number of counters is ten. Therefore, the new *whole* totals ten. How many counters must be removed from this group to take it back to its original amount? Two. What fraction of the new total is this? One-fifth. What percentage is this? It is 20 per cent (see diagram).

8 counters

Increase by 25%

New total 10 counters

Remove one-fifth (20%) of the counters

From this demonstration, we see that if we increase something by 25 per cent, we actually need to reduce it by 20 per cent to get back to the original amount. Relating this back to the shop owner, a 25 per cent discount after increasing the original price by 25 per cent means that the shop owner is selling the product for less than the price before the sale. The shop owner would not have a hope of breaking even or making a profit.

Through modelling with counters, the following relationships can be discovered:

- An increase of 50 per cent means a reduction of 3313 per cent to get back to the original.
- An increase of 25 per cent means a reduction of 20 per cent to get back to the original.
- An increase of 3313 per cent means a reduction of 25 per cent to get back to the original.
- An increase of 100 per cent means a reduction of 50 per cent to get back to the original.

Teaching rational number in the middle years

Rational number is situated squarely in the middle years mathematics curriculum. In fact, the focus of number in the middle years is on promoting rational number knowledge. In the early middle years, decimal and common fraction understanding is the focus. This is extended gradually to ratio and percentage as students progress to the upper middle years. The study of rational number is difficult for students, and students will exhibit varying levels of understanding in the middle years of schooling. In this chapter, foundations for rational number have been provided, which enables teachers to look back to activities for building conceptual understanding when students experience difficulty. Using concrete materials and structuring learning experiences will support students at any year level, and these strategies have been advocated in this chapter.

Summary

For many students, rational number topics pose great difficulty. It is important to provide opportunities for students to engage with the big ideas of rational number topics so that they can both see connections between these topics and also understand key ideas within each topic. Concrete materials need to be carefully selected on the basis of their value in accurately representing key concepts for rational number understanding, and the associated language must become a natural part of the students' dialogue. Often confusion with rational number topics occurs as a result of rushing to symbolic representation and manipulation of numbers before a rich conceptual framework has been developed sufficiently. Rational number study needs to be an integral part of early mathematics classrooms in an exploratory sense so that a solid foundation is laid for more formal investigations in the middle and later years.

■ REVIEW QUESTIONS

11.1 Describe rational numbers in terms of their location on a number line. Outline how rational numbers extend students' conceptualisation of numbers on a number line.

11.2 Why is decimal knowledge only partly dependent upon fraction knowledge?

11.3 Describe how a fraction meaning may interfere with developing ratio meaning.

11.4 Outline the multiplicative structure of ratio and percentage increase situations.

Further reading

Condon, C. and Hilton, S. (1999). Decimal dilemmas. *Australian Primary Mathematics Classroom*, 4(3), 26–31.

Cramer, K. and Bezuk, N. (1991). Multiplication of fractions: Teaching for understanding. *Arithmetic Teacher*, 39(3), 34–7.

Empson, S. (2001). Equal sharing and the roots of fraction equivalence. *Teaching Children Mathematics*, 7(7), 421–5.

Graeber, A.O. and Baker, K.M. (1992). Little into big is the way it always is. *Arithmetic Teacher*, 39(8), 18–21.

Haubner, M.A. (1992). Percents: Developing meaning through models. *Arithmetic Teacher*, 40(4), 232–4.

Mack, N.K. (1998). Building a foundation for understanding the multiplication of fractions. *Teaching Children Mathematics*, 5(1), 34–8.

Moloney, K. and Stacey, K. (1996). Understanding decimals. *The Australian Mathematics Teacher*, 52(1), 4–8.

Pothier, Y. and Sawada, D. (1990). Partitioning: An approach to fractions. *Arithmetic Teacher*, 38(9), 22–4.

Steffe, L.P. and Olive, J. (1991). The problem of fractions in the elementary school. *Arithmetic Teacher*, 38(9), 22–4.

Thompson, C.S. and Walker, V. (1996). Connecting decimals and other mathematical content. *Teaching Children Mathematics*, 2(8), 496–502.

Witherspoon, M. L. (1993). Fractions: In search of meaning. *Arithmetic Teacher*, 40(8), 482–5.

Yang, D. and Reys, R. (2001). One fraction problem, many solution paths. *Mathematics Teaching in the Middle School*, 7(3), 164–6.

References

Ashlock, R.B. (1994). *Error patterns in computation: A semi-programmed approach* (6th ed.). New York: Macmillan.

Dole, S. (1999). Successful percent problem solving for Year 8 students using the proportional number line method. In J.M. Truran and K.M. Truran (eds), *Making the difference: Proceedings of the 22nd Annual Conference of the Mathematics Education Research Group of Australasia*. Adelaide: MERGA.

——(2000). Promoting percent as a proportion in eighth-grade mathematics. *School Science and Mathematics*, 100(7), 380–9.

Dole, S., Cooper, T., Baturo, A. and Canoplia, Z. (1997). Year 8, 9 and 10 students' understanding and access of percent knowledge. In F. Biddulph and K. Carr (eds), *People in mathematics education: Proceedings of the 20th annual conference of the Mathematics Education Research Group of Australasia*. University of Waikato Printery, NZ: MERGA.

Glatzer, D. J. (1984). Teaching percentage: Ideas and suggestions. *Arithmetic Teacher*, 31, 24–6.

Lesh, R., Post, T. and Behr, M. (1988). Proportional reasoning. In J. Hiebert and M. Behr (eds), *Number concepts and operations in the middle grades* (pp. 93–118). Hillsdale, NJ: Lawrence Erlbaum.

Middleton, J.A. and van den Heuvel-Panhuizen, M. (1995). The ratio table. *Mathematics Teaching in the Middle School*, 1(4), 282–8.

Ministry of Education, New Zealand. (2008). *The Number Framework*. Wellington: Ministry of Education.

Parker, M. and Leinhardt, G. (1995). Percent: A privileged proportion. *Review of Educational Research*, 65(4), 421–81.

Post, T. R., Behr, M.J. and Lesh, R. (1988). Proportionality and the development of prealgebra understandings. In A.F. Coxford and A.P. Shulte (eds), *The ideas of algebra, K–12* (pp. 78–90). Reston, VA: NCTM.

Stacey, K. and Steinle, V. (1998). Refining the classification of students' interpretations of decimal notation. *Hiroshima Journal of Mathematics Education*, 6, 1–21.

CHAPTER 12

Patterns and algebra

For many students, the study of algebra is associated with a turning point—or, more precisely, a turning-off point—in the study of school mathematics. They leave school with vague feelings of uneasiness associated with *x* and *y*, and trying to make sense of something they feel is quite meaningless. Yet this should not be the case, as algebra understanding and algebraic thinking lead on from early learning about numbers and patterns. 'Difficulties that students experience in algebra are not so much difficulties in algebra itself as problems in arithmetic that remain uncorrected.' (Booth, 1988).

Algebra is often referred to as generalised arithmetic, where students take their early learning about number and operations, and move to seeing generalisations. Exploring patterns is key. Patterning has long been a main element of the mathematics curriculum in the early years, and its importance as a foundation for algebra is now increasingly being recognised and acknowledged (Usiskin, 1997; Carraher & Schliemann, 2007). 'Algebra learning has its roots in the early grades, when children notice regularities in the ways numbers work.' (MacGregor and Stacey, 1999)

For many students, their introduction to formal algebra is through endless textbook exercises involving solving equations and inequalities, finding the value of *x*, collecting like terms and substituting values for unknowns. Although this is part of what comprises algebra knowledge

(Kieran, 2007), if taught without meaning, there is little chance that students will see algebra as useful. Algebra assists in solving real problems, and therefore is a valuable numeracy tool.

Learning algebra is important from a perspective of equity, for algebra is regarded as a key to mathematical literacy, which in turn is the key to economic access and citizenship (Bond, 2003). Providing activities that involve simple patterning involves students thinking logically, and this in turn supports their general learning of mathematics. Algebraic thinking is supported through the work of patterning and relationships, thus mathematics education in the primary school needs to have a strong emphasis on patterning in order to prepare students for algebraic thinking. For many students in the middle years of schooling, experiences with algebra are problematic because they have not had a good foundation in early algebra.

Patterning

There are three distinct aspects of the teaching of patterning: repeating patterns, growing patterns and number patterns. Early experiences should be ones that students can easily understand, and teachers should organise learning experiences that progressively extend students' thinking. It is important to allow students time to justify, explain or extend their patterns since there may be other ways to see a pattern than that envisaged by the teacher. Observing students working with patterns can provide a basis for understanding their thinking, and so allow the teacher to better plan for teaching. Blume and Heckman (2000) note that students are more likely to be able to continue a pattern than to insert a missing element. It is important for students to experience both linear patterns—that is, patterns in one direction only, as well as patterns in two directions—such as those in the picture.

When teaching patterning, it is important to focus on students developing appropriate language relating to patterns so that they become

Early experiences with patterning in two directions

familiar with identifying particular elements within a pattern—for example, what will the fifth element be?

● Repeating patterns

Repeating patterns are usually introduced in a linear form. Once students are familiar with this format, they can work with formats where the repetition goes in more than one direction. Study of patterns can often involve geometric patterns and patterns in cultural activities, such as quilting, mat weaving and so on.

Early pattern work will involve patterns of various colours or shapes. As a general rule, early work should involve patterning where the only attribute is the focus of the pattern. Blocks that vary *only* in colour *or* shape should be used for early work. Use of Unifix blocks can be a good starting point since these blocks are identical except for colour. This allows students to focus on the changes in colour. Experiences with patterning involve students:

- copying a pattern;
- describing a pattern;
- copying and extending a pattern; and
- making and describing a pattern.

The cognitive demands placed on students with these patterning activities increase, so early work should involve considerable copying of patterns and building towards creating patterns. Patterns may initially involve only one of each item (red, green, blue, red, green, blue) but later will involve multiples of an item (red, red, green, red, red, green).

Early experiences with patterns should consist of only two items in one attribute. These are commonly referred to as 'AB patterns'. Typically, this could be of a shape (square, circle, square, circle) or other attributes, as shown in the diagram. Various items commonly found in classrooms can be used for this type of patterning.

Early experiences in patterns can vary in one attribute such as colour, shape or size, as in the next diagram. The shape and size patterns here are simple ABC patterns. As students become

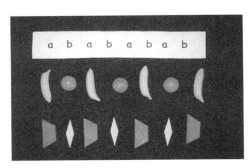

Simple AB patterns

confident in continuing or creating patterns of this format, activities can focus on making the patterns more complex by using formats such as ABB or ABBC. While geometric shapes are often the basis of early patterning, activities can also involve the use of alphanumeric-symbolic forms, pictures, and so on.

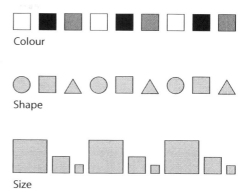

Once students are familiar with repeating patterns where there is a change in one attribute, teachers can introduce changes to the other attributes. In the example below, both colour and shape have been changed to create a more complex pattern which includes first-order and second-order patterns—that is, a smaller pattern is contained within a larger pattern. Here, the first-order pattern is related to the shapes—square, square, circle; this is contained within a larger pattern of colour—black, grey, white, shade. The first-order pattern of shape is contained within

the second-order pattern of colour. This is fairly complex and would be introduced later in the process of patterning.

● Growing patterns

Rather than a sequence of patterns that is repeated over and over again, a growing pattern involves an increase or growth in the subsequent sections of the sequence when compared with the previous sections. An example of a growing pattern is AB, AAB, AAAB. Here, the number of Bs remains unchanged but the number of As increases by one in each subsequent section. While there is a change in each section, the change is constant (increases by one)—that is, the second element in the pattern is growing (or increasing) by one each time while the first element repeats. Initial activities should involve only one element of the pattern increasing, but as students become confident with growth in one element, growth in more than one element can occur.

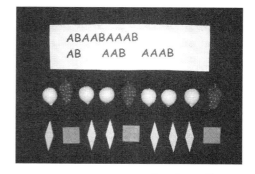

Growing patterns

Growing patterns can also be two dimensional. This can involve the use of tiles so that the pattern can be seen to grow in both directions.

● Number patterns

Number patterns, like growing patterns, are about describing changes. For example, in the number sequence 2, 4, 6, what number comes next? Students need to identify the change that is occurring (adding 2 here), or in another example, notice that when subsequent numbers are compared to the first, the second has increased by 2, the third by 4, so that the next will increase by 6; thus the number after 6 is 8.

Students should come to realise that when working with patterns in sequences of numbers, knowing the first three terms only might not be sufficient to develop a theory to explain the pattern. In some cases, more than three terms might be required. Also, as indicated in the picture, there might be several ways to describe a particular pattern.

Simple sequences of numbers

Students should be provided with a variety of activities such as finding the next three terms in the sequence 2, 4, 6, 8 . . . , or finding terms at the beginning of a sequence such as . . . 31, 38, 45, 52. Students should be encouraged to describe in words the sequence and the change from one term to the next. The teacher's role is to explicitly draw students' attention to making generalisations about number patterns. When students are adept at counting in twos, ask them to look at the relationship between the number and its solution:

- What is the first number? Two.
- What is the second number? Four.
- What is the third number? Six.
- What do you notice about the position of the number in the sequence and its solution? It is double.
- So . . . what is the tenth number in the sequence? Twenty.
- What is the hundredth number? Two hundred.

In simple algebraic terms, this means that the general rule is $2n$. This is actually an exploration of even numbers (see below).

● Odd and even numbers

The earliest experiences with number theory arise through the study of odd and even numbers. This is often taught through the use of grids (or egg cartons) where students can see what happens when odd and even numbers of items are placed into a two-grid (see diagram).

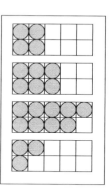

Opportunities should be provided for students to develop their own theories about odd and even numbers, to articulate what they see and to develop a sense of what is happening. They will also need opportunities to determine for themselves the following patterns: even numbers end with 0, 2, 4, 6 or 8 and odd numbers end in 1, 3, 5, 7 or 9. Like developing the generalisation that $2n$ is the algebraic shorthand for even numbers, a generalisation for odd numbers can occur by drawing students' attention to the position in the sequence and its solution:

- What is the first number? One.
- What is the second number? Three.
- What is the third number? Five.
- What do you notice about the position of the number in the sequence and its solution? It is double minus 1.
- So . . . what is the tenth number in the sequence? Double 10 minus 1, which is 19.
- What is the 100th number? It is 199.

In algebraic terms, this means that the general rule for an odd number is $2n - 1$. As with all scaffolding, good questions are important for students to develop robust understandings. Another good question to pose is whether zero is even or odd.

Rich discussion can ensue from such questions. By the way, zero is an even number. The definition of an even number is that it is exactly divisible by 2. Multiplying and dividing numbers by 2 and exploring patterns will lead to healthy discussion on why zero is even and, further, whether odd and even numbers can be negative numbers (and yes, they can).

● Patterns involving two related sequences of numbers (relationships)

Mathematics can involve finding and describing relationships between items. Early work with relationships focuses on simple experiences— such as how the number of bicycles is related to the number of wheels; or how the number of hands is related to the number of fingers. The study of relationships should link closely to other aspects of number study—for example, it should involve the operations $(+, -, \times, \div)$ currently being studied, and numbers in the range currently being studied.

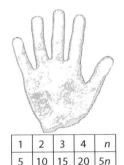

| 1 | 2 | 3 | 4 | n |
| 5 | 10 | 15 | 20 | $5n$ |

At some point, students need to develop effective ways to record their work. Tables or grids can be used as shown in the diagram. Terms in the upper row increase by 1, and students determine the constant increase in the lower row (in this case, 5). A more complex response is to answer in the form of a generalisation—that is, to describe a rule for determining terms in the lower row from the corresponding term above. In this case, the generalisation is that the number is multiplied by 5 ($5n$). By developing a rule in this way, students can predict any term in the sequence. Consider the task of working out the one-hundredth term in the sequence. This is made much easier by using the generalisation or rule.

● Function machines

'Guess my rule' is an activity used with the aid of a function machine. To construct a function machine, simply draw a square shape on the board, with a 'funnel' shape going into the square on one side and a funnel coming out of the square on the other side. Think of a rule that can be applied to a number (e.g. double the number, then add 1). Students volunteer a number, and it is written on the board near the incoming funnel of the function machine. Once the rule has been applied, the outgoing number is written near the outgoing funnel of the function machine. For example, 7 goes in, 15 comes out. After several numbers going in and coming out, ask students to volunteer what they think the rule is. There are many interactive websites that contain animated function machines, which can be used with the whole class via an interactive whiteboard, or at single computers for small-group or individual work. Just google function machines and browse the range of options available. An even easier approach is to provide students with simple tables of values, with columns headed 'input' and 'output' for exploration and analysis. Ensure students also have the opportunity to make up their own rules for the whole class to determine.

Function machines and the types of pattern exploration described above can provide a context for introducing students to graphing (or plotting) pairs of numbers on a grid. In this way, students can be introduced to displaying the function rule geometrically. For example, consider the simple function machine in the diagram, where the function (or change) is add 3. In this example, a number—say, 5 (the input)—comes into the function machine, the change (+ 3) is made and the number 8 (the output) exits from the machine. A table showing input numbers 1 to 6, and the corresponding output numbers, can then be constructed:

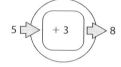

+ 3	1	2	3	4	5	6
	4	5	6	7	8	9

The numbers can then be plotted on a grid showing the horizontal numbers for 1 to 6 and the vertical numbers to 9. When the numbers are plotted, a straight line appears.

The plotted line (graph) can then be used to determine other output values for given input values. For example, if the input is 7, what is the output? If the graph is extended over a longer range of values, other output numbers can be read off. Students can also work in this way with numbers involving decimals (or fractions). For example, what

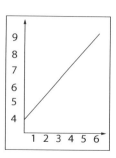

input would give an output of 7.5? The student's task is to determine this from the graph.

Using approaches such as these, teachers can informally introduce ideas and language relating to algebra and functions. Providing students with variety in representing patterns and relationships to include tables, graphs, words and symbols is an important part of algebra knowledge (Kieran, 2007).

● Technology and algebraic thinking

There are many interesting ways in which computers can be used to support the development of students' mathematical thinking. One that is very relevant to number patterns and algebra involves the use of spreadsheet software (Ploger et al., 1997). A spreadsheet is a table of numbers arranged in rows and columns. These days, every common computer is equipped with spreadsheet software—that is, a program that will do calculations automatically according to formulae entered into the computer. Thus, in contexts such as the study of interest rates, foreign currencies or results of sports competitions, students can use spreadsheets to do the routine aspects of calculation, and thereby focus on thinking about the results of the calculations. Spreadsheets are also an ideal tool for undertaking activities involving function machines or related sequences of numbers (see above).

One type of activity can involve students reflecting on two related columns of numbers in order to determine a number pattern and to study or work out the formula involved. This process encourages students to develop a general rule, articulate it in a language applicable to the software and then carry out the necessary tasks to program the spreadsheet. Most spreadsheet programs are simple to use and most are appropriate for upper primary students. For example, if students were converting foreign currencies into the local currency, they could fill in the first cell, work out the multiplier for the second cell and then, using the appropriate spreadsheet language, fill in the commands for the program being used and, using the 'fill' mechanism, complete the third cell. This process encourages students to think algebraically in terms of developing and articulating the general rule needed.

Number, arithmetic and algebra

Through number study in the early years, students are working with powerful algebraic ideas, but often aren't aware of it. When learning basic facts, for example, students are introduced to the commutative law of arithmetic—when they learn one basic fact, they are actually learning two as they see 7×5 is the same as 5×7. When they are performing mental computation, the associative law of arithmetic is often used. For example, when adding $9 + 7 + 3$, it is probably easier to add 7 and 3 first, then add 9, rather than begin at the start and add 9 and 7, then 3. These understandings about numbers and how they work are important foundations for algebra. This does not mean that explicit teaching of the laws of arithmetic is required, but rather that students need to be immersed in an environment in which rich number explorations and investigations are a feature of number study.

● Equality

Early algebra research identified that students' conception or misconception of the equals sign was actually a barrier to algebraic thinking. Early experiences of number sentences such as $3 + 4 = 7$ is said as '3 and 4 makes 7'. The concept of 'making' results in confusion because it does not necessarily highlight the idea of equivalence (Falkner et al., 1999; Warren and Cooper, 2003). For example, when posing the question: 'Is this sentence correct: $4 + 5 - 2 = 7$?', students might regard it as incorrect since it does not conform to their expectations. Commonly, $5 - 2 = 3$ is regarded as back to front, or it is believed that the answer is 5. Students are not aware of the relationship between 5 and $2 + 3$, or that this involves equality. Such misunderstandings are a result of narrow teaching and lack of opportunity to explore numbers and the ways in which they interrelate. Balance scales, or a number beam balance, can be used to convey the idea of the equals sign as representing equivalence. Students place blocks on each side of the beam to make it balance. They then remove blocks, realising that they must remove an equivalent number of blocks from the other side to keep the beam in balance. This supports later algebraic activities of solving for unknowns. As with all manipulatives, the value in the balance beam relates to how it is used and whether students are able to abstract the notion of equality from the balancing process.

● Arithmetic number study

Some important terms associated with the four basic operations of arithmetic are presented in the table below.

Operation	Key terms	Example	Application of terms
Addition	Addend, sum	$6 + 4 = 10$	Addends: 6 and 4 Sum: 10
Subtraction	Subtrahend, difference, minuend	$15 - 12 = 3$	Subtrahend: 12 Difference: 3 Minuend: 15
Multiplication	Factor, product (multiplier, multiplicand)	$8 \times 4 = 32$	Factors: 8 and 4 Product: 32 Multiplier: 4 Multiplicand: 8
Division	Divisor, dividend, quotient, remainder	$43 \div 8 = 5\,r\,3$	Divisor: 8 Dividend: 43 Quotient: 5 Remainder: 3

● Laws of arithmetic

A mental computation program assists students to become familiar with the laws of arithmetic. Exploring strategies for mental computation supports flexible thinking about number and understanding of number relationships. The associative law is learnt through spin-arounds, when learning basic facts ($9 + 3 = 3 + 9$). Students are meeting the inverse law when they learn subtraction and division facts. Once addition and multiplication facts are known, students develop subtraction and division facts by using 'think addition' and 'think multiplication' respectively. When students learn addition of zero or multiplication by

1, they are learning the identity law. The laws of arithmetic are provided in the table below.

Law	Addition	Multiplication
Commutative	$a + b = b + a$ $(9 + 3 = 3 + 9)$	$a \times b = b \times a$ $(9 \times 3 = 3 \times 9)$
Associative	$(a + b) + c = a + (b + c)$ $(9 + 7) + 3 = 9 + (7 + 3)$	$(a \times b) \times c = a \times (b \times c)$ $(4 \times 7) \times 5 = 4 \times (7 \times 5)$
Distributive	$a \times (b + c) = a \times b + a \times c$ $74 \times 98 = 74 \times 100 - 74 \times 2$	$(a + b) \times c = a \times c + b \times c$ $102 \times 56 = 100 \times 56 + 2 \times 56$
Identity	$a + 0 = a$ $5 + 0 = 5$	$a \times 1 = a$ $6 \times 1 = 6$
Inverse	$a + (-a) = 0$ $3 - 3 = 0$	$a \times \frac{1}{a} = 1$ $4 \times \frac{1}{4} = 1$

Factors and multiples

A factor is a whole number that can be divided equally into another number. A multiple is a whole number that is created by multiplying the original number by a whole number. A good way to discuss some of the language of factors and multiples is through the use of examples: 6 is a multiple of 2; 20 is a multiple of 5; 100 is a multiple of 10. The general rule for multiples is that a is a multiple of b if a can be obtained by multiplying b by a whole number. Factors can be explored in a similar way: 4 is a factor of 12, 8 is a factor of 16, and so on. The general rule for factors is that a is a factor of b if b can be obtained by multiplying a by a whole number.

Prime and composite numbers

Through the study of factors and multiples, students come to see that some numbers in the 1 to 100 range have a relatively large number of factors. For example, 36 has nine factors—1, 2, 3, 4, 6, 9, 12, 18 and 36; and 90 has 12 factors—1, 2, 3, 5, 6, 9, 10, 15, 18, 30, 45 and 90. They will also discover that there are some numbers that have no more than two factors—that is, 1 and the number itself. For example, 13 has 1 and 13 as its only factors. Numbers with two factors only are referred to as *prime numbers*; numbers with more than two factors are referred to as *composite numbers*.

● Number explorations

Geometric numbers

Exploring geometric numbers and trying to find the general way to describe the pattern is another way to promote algebraic reasoning and flexible number knowledge. Using concrete materials and constructing tables of values are the starting points. Use counters to create square numbers, triangular numbers and rectangular numbers, and keep a record (using an organised list or table) of the number of counters required to construct each number in the sequence. Look for a relationship between the position of the number and the number of counters required. Try to determine the tenth, then the hundredth number in each sequence.

The table below shows the number of counters required to construct the first five square, triangular and rectanglular numbers.

Position	1	2	3	4	5
Square number	1	4	9	16	25
Triangular number	1	3	6	10	15
Rectangular number	2	6	12	20	30

Palindromic numbers

Palindromic numbers are numbers that read the same from left to right or right to left. Examples of these are 121, 66, 147 741. Students should be familiar with palindromes from their language studies, and are often excited by seeing numbers working in a similar way.

Palindromic numbers can be obtained through an activity where a number is written down, and the reverse of that number is written below it. The numbers are then added together. If their sum is a palindromic number, it is said to be a first-order palindromic number. If the result is not a palindromic number, the cycle is repeated until a palindromic number is obtained. For example, starting with 728, this is reversed to 827 and the two numbers are added. The sum is 1555 (which is not palindromic), so the cycle is repeated: 5551 + 1555 = 7106, which again is not palindromic. The third cycle results in 13 123, the fourth in 45 254, which is a palindromic number. For any number, repeating this cycle of operations will eventually produce a palindromic number.

The importance of algebra in the primary school

MacGregor and Stacey (1999) argue that the roots of algebra learning arise in mathematics in the early grades when students begin to notice regularities in calculations with numbers. They discuss five aspects of number work in the primary grades that are important as a basis for algebra: understanding equality; using a wide range of numbers; understanding important properties of numbers; recognising the operations; and describing patterns and functions.

Understanding equality relates to understanding number sentences and algebraic equations, and in particular the meaning of the equal sign in number sentences and algebraic equations (Warren and Cooper, 2003). Using a wide range of numbers relates to the importance of students working with fractions and decimals, as well as with whole numbers, because numbers in the forms of fractions and decimals can occur frequently in algebra.

An example of an important property of numbers which needs to be understood as a basis for algebra is the following: if you subtract one less from a number then the answer is one more. For example, $2475 - (357 - 1) = (2475 - 357) + 1$. Understandings like this are necessary when algebraic equations are used to solve problems.

Recognising the operations relates to the need for students to develop a full understanding of addition, subtraction, multiplication and division as a basis for using algebra in problem-solving. This is because algebra focuses explicitly on *properties* of operations, and on *relations* among operations, rather than merely *carrying out* operations, which is the main focus in arithmetic. Important examples of properties of operations (i.e. addition, subtraction, multiplication and division) are the commutative, associative and distributive properties (also known as 'principles' or 'laws'). Algebra at the primary school level does not include a formal study of these properties. Rather, the focus is on developing intuitive understandings of these properties in working with numbers. In relation to patterns and functions, the focus of this chapter has been the importance of opportunities for students to explore patterns, find general rules and make generalisations.

Teaching algebra in the middle years

For many students, their formal introduction to algebra topics is in the middle years when they start secondary school mathematics. And for many students, this formal introduction to algebra is the turning point, at which mathematics no longer makes sense. To support students in the middle grades, early patterning experiences in geometry and number (as described in this chapter) are essential. In the middle years, the task is to build on this knowledge to support meaningful understanding of symbol-manipulation exercises and other algebraic activities.

Algebra is a major tool for supporting problem-solving in the real world. Consider Brian the hot-dog seller. Brian is trying to make money to help pay for uni by selling hot dogs from a cart at the football. He pays the owner of the cart \$35 per game for the use of the cart. He sells hot dogs for \$1.25 each. His costs for the hot dogs, condiments, serviettes and other paper products are about 60 cents per hot dog on average. Brian can readily calculate the number of hot dogs he needs to sell each day to break even once he has constructed an equation for the situation ($0.65x-\$35$). Providing students with real-world problems such as this will support their capacity to use and apply algebra in their lives, and support the development of beliefs about the value of mathematics.

■ REVIEW QUESTIONS

12.1 Outline the process you might undertake to teach patterning to students.

12.2 In your staffroom, there has been heated debate about the place of algebra in the primary school. Outline the position that you would take in relation to teaching algebra and justify your position.

12.3 Equality can often be a difficult concept for students to grasp. Outline some of the ways in which current teaching methods contribute to the difficulties encountered by students, and devise three strategies you could use to support students to develop this important mathematical idea.

12.4 Outline some approaches to teaching students about the commutative property in the contexts of addition, subtraction, multiplication and division. Some of your approaches might involve the use of concrete materials.

12.5 Determine some approaches to teaching about prime, composite, square and triangular numbers using two-dimensional shapes.

12.6 Technology is important in the development of algebraic thinking. Devise some strategies that could be implemented in the classroom.

Further reading

Bay-Williams, J. (2001). What is algebra in elementary school? *Teaching Children Mathematics*, 8(4), 196–200.

Bishop, J., Otto, A. and Lubinski, C. (2001). Promoting algebraic reasoning using students' thinking. *Mathematics Teaching in the Middle School*, 6(9), 508–14.

Charters, M. (1996). Visualising: Playing powerfully with numbers. *Australian Primary Mathematics Classroom*, 1(1), 22–7.

Clements, D.H. and Sarama, J. (1997). Computers support algebraic thinking. *Teaching Children Mathematics*, 3(6), 320–5.

Fenton, P. and Watson, J. (2001). Triangular numbers: Facts or fiction? *Australian Primary Mathematics Classroom*, 6(1), 10–14.

Norman, F.A. (1991). Figurate numbers in the classroom. *Arithmetic Teacher*, 38(7), 42–5.

Vogel Boyd, B. (1987). Learning about odd and even numbers. *Arithmetic Teacher*, 35(3), 18–21.

Yolles, A. (2001). Making connections with prime numbers. *Mathematics Teaching in the Middle School*, 7(2), 84–6.

References

Blume, G. W. and Heckman, P.E. (2000). Algebra and functions. In E. Silver and M.J. Kenney (eds), *Results of the 7th mathematics assessment of the national assessment of education progress*. Reston VA: NCTM.

Bond, M. (2003). Powers to the people: Interview with Robert Moses. *New Scientist,* 19 April.

Booth, L. (1988). Children's difficulty in beginning algebra. In A.F. Coxford and A.P. Shulte (eds), *The ideas of algebra K–12: 1988 Yearbook* (pp. 20–32). Reston, VA: NCTM.

Carraher, D. and Schliemann, A. (2007). Early algebra and algebraic reasoning. In F. Lester (ed.), *Second handbook of research in mathematics teaching and learning* (pp. 669–706). Charlotte, NC: Information Age.

Curio, F.R. and Schwartz, S.L. (1997). What does algebraic thinking look like and sound like with preprimary children? *Teaching Children Mathematics*, 3(6), 296–300.

Falkner, K., Levi, L. and Carpenter, T. (1999). Children's understanding of equality: A foundation for algebra. *Teaching Children Mathematics*, 6(4), 232–6.

Ferrini-Mundy, J., Lappan, G. and Phillips, E. (1997). Experiences with patterning: An example for developing algebraic thinking. *Teaching Children Mathematics*, 3(6), 282–8.

Kieran, C. (2007). Learning and teaching algebra in the middle school through college levels: Building meaning for symbols and their manipulation. In F. Lester (ed.), *Second handbook of research in mathematics teaching and learning* (pp. 707–804). Charlotte, NC: Information Age.

MacGregor, M. and Stacey, K. (1999). A flying start to algebra. *Teaching Children Mathematics*, 6(2), 78–85.

Ploger, D., Klingler, L. and Rooney, M. (1997). Spreadsheets, patterns and algebraic thinking. *Teaching Children Mathematics*, 3(6), 330–4.

Usiskin, Z. (1997). Doing algebra in grades K–4. *Teaching Children Mathematics*, 3(6), 346–56.

Warren, E. and Cooper, T. (2003). Introducing equivalence and inequivalence in Year 2. *Australian Primary Mathematics Classroom*, 8(1), 4–8.

CHAPTER 13

Measurement

What is measurement?

The measurement strand involves the assignment of a numerical value to an attribute of an object or event. While there is a degree of arbitrariness to what is included in this content area and the way it is organised (whether around content or processes), the focus of the strand is the measurement of some attribute. In some cases, probability is seen as the measurement of chance and will be assigned to this strand, whereas in other documents it will remain as its own strand. Similarly, measuring angle is the measurement of the turning attribute and could be part of this strand. However, in some documents it is seen to be an integral part of geometry and located in the space strand.

Measurement has an important role to play in the mathematics curriculum. The National Council of Teachers of Mathematics (NCTM, 1989) notes that 'measurement is of central importance to the curriculum because of its power to help children see that mathematics is useful in everyday life and to help them develop many mathematical concepts and skills' (p. 51). The NCTM further argues that measurement pervades 'so many aspects of everyday life . . . [It] offers the opportunity for learning and applying other mathematics, including number operations, geometric ideas, statistical concepts and notions of function. It highlights connections within mathematics and between mathematics and areas outside of mathematics, such as social studies, science, art and physical education.' (NCTM, 2000, p. 44).

To consider the importance of measurement, stop and think about when you have used mathematics in the past week. You will notice that it was usually in relation to measurement—how far you walked, how much money you spent, how many drinks you had during the day, how much petrol was needed to fill the car. As measurement tends to be the strand of mathematics that is used most in everyday life, measurement should be taught in a very practical manner, making it highly enjoyable for students (and teachers).

For teachers, the measurement strand is useful for linking across other curriculum areas, thus making its teaching highly integrative. It links strongly with the number strand (calculating area involves multiplication); using analogue clocks reinforces fraction work; conversion of units of measures uses number work, and so on. Similarly, links can be made across curriculum areas—timing sporting events such as running (time); international travel in social studies (timelines, geometry on a sphere); writing reports for science and English (using measurements of temperature, calculations, representation of data). Similarly, within mathematics many of the topics that are taught in other strands have a strong connection with measurement. For example, the use of the number line links in with length; multiplication grids link in with rectangular areas; measuring to the nearest unit involves rounding.

Aside from the mathematical aspects, one of the more powerful aspects of teaching the measurement strand is the pedagogical benefits. As the teaching of measurement frequently involves hands-on work, students enjoy the change of pace in lessons. Often students do not even see that what is being taught is mathematics (particularly where they might have a perception of mathematics as being a boring, solitary activity, involving things like times tables). Organising a class party where they have to budget (money, operations), work out best buys (ratio), cook food (volume, mass), shop (money), fund-raise and plan (working mathematically) creates a very positive atmosphere in the classroom as the students work collaboratively on realistic tasks with meaningful ends. The social dynamics of the class improve, the students enjoy the tasks and they learn considerable mathematics without even being conscious of it. Measurement gives students the opportunity to apply their knowledge and skills in a meaningful and enjoyable manner.

In summary, the measurement strand is an integral part of the mathematics curriculum for several reasons including the following:

- Measurement applies to the everyday experiences of students.
- Measurement provides application to the world beyond school.
- Measurement supports other areas of the mathematics curriculum.
- Measurement supports other areas of the curriculum.
- Measurement involves active, hands-on learning experiences.
- Measurement strongly links to problem-solving.

● Sub-strands of measurement

In most curriculum documents, the recurring themes centre around eight common concepts: length, area, volume, mass, time, temperature, money and angle. There is variation, as noted earlier, in how and where these sub-strands are located in any particular curriculum.

Length

Length is the most easily perceived of the measurement attributes, and refers to how long or how far something is. This length can be along a single plane or through two dimensions (such as around the circumference of a bottle). 'Length' refers to the attribute, while other terms such as 'width' or 'height' refer to specific contexts for length. In most cases, students come to school with some concept of length and the language associated with it. Many students have misconceptions of length, such as those illustrated with Piaget's conservation of length tasks, but as they progress through school and their lives beyond school, their experiences in most cases allow them to construct concepts that are more rigid. Frequently, length is seen to be an attribute of something long and thin, like a piece of string or a line, but over time this conception gives way to other conceptions, including the distance between two points—for example, home and school—or the circumference of circular objects such as bottles or cans. Distance becomes another aspect of length; it is often more difficult for students, so it tends to be introduced later in schooling. Distance has a specific language (such as near and far).

"Close enough??"

Area

Area is a two-dimensional concept that is related to the region enclosed within a plane shape. Students usually begin exploring area by investigating the amount of surface area. They do this through a process of 'covering'. Teaching methods in the past frequently relied on teaching rote formulae ($A = L \times W$), but these resulted in many students being confused between area and perimeter. It is now recognised that students should be taught so that they have an understanding of the concept and develop intuitive ways to calculate area so that formulae become simply a formal recognition of the work they have progressively been developing.

Volume and capacity

Volume is the amount of space taken up by an object. Capacity is the amount an object can hold. Often, but not always, volume can be seen as a solid measure and capacity as a liquid (or fluid) measure. By working in this sub-strand, students learn the relationship between millilitres and litres, and between cubic centimetres and cubic metres.

Mass

Mass and weight are often confused: while they are very different mathematically and scientifically, for most everyday experiences they are the same, so it is not uncommon for the terms to be used interchangeably (even by teachers). It is most common to hear people referring to the process of 'weighing' when they are in fact measuring mass. Mass refers to the amount of inertia required to stop or start an object into motion. It is often identified through the process of hefting. For example, it requires far greater force to move a greater mass than a smaller mass. Similarly, if the same amount of force is applied to two objects of differing mass, the smaller object will move further than the larger one. In contrast, weight is the force that gravity exerts on an object. Thus, while mass will remain constant, weight can vary depending on where an object is located. The weight of an object is less on the moon than it is on earth.

Angle

Angle refers to the amount of turn or rotation. Most frequently, it is taught as a static measure such as the distance between two rays emanating from the same point (see diagram A). It is often seen in activities where students are expected to compare the sizes of two-dimensional corners in a shape. However, mathematically it is more correct to refer to the amount of rotation. Because it has been taught as the 'area' between two rays and measured with a protractor, students frequently construct the misconception that angles are less than 180°, so that when they are required to measure rotations greater than 180° they become confused, as they do not see this as an angle (see diagram B). This sub-strand is often located within the geometry or space strands of the curriculum.

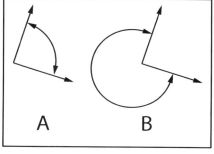

A   B

Interior angle   Exterior angle

Temperature

Temperature is used to measure how hot or cold things are. In most cases, the difference between hot and cold is apparent. Certain circumstances can cause misperceptions of temperature to occur, creating the need for formal measurement with a thermometer. Different scales have been designed to measure temperature, where arbitrary units have been assigned to key experiences—water boiling and freezing. The Celsius system uses the arbitrary measures of zero for water freezing and 100° for water boiling. Other systems use different scales for these key points: Fahrenheit, for example, uses 32° for freezing and 212° for boiling. The Celsius system has created a need for negative numbers to identify temperatures below freezing. Depending on students' out-of-school experiences, temperature can provide the ideal learning situation for the introduction to negative (or directed) numbers. It can also be used for linking with other curriculum areas (Moore, 1999).

Value (money)

Often the sub-strand of money results in the misperception that money itself is the item of measurement. More correctly, money is the unit

of measure of a particular concept, in this case value (or cost). In most Western countries, money is used to measure the value or cost of something. Unlike other measures, value can be very flexible, as can be seen in the fluctuations in currency exchange rates. The cost of an item may not always represent its value. For example, people are often prepared to pay more for something they particularly want (such as a house at auction, or collector cards) than it would have been worth to them had they not been so keen to purchase.

Money is frequently found in the measurement strand, although in some curricula it is found in the number strand. The justification for inclusion in the number strand is money's close relationship to number, particularly decimals. Students in many countries encounter money as whole dollars and part cents, which means that they encounter decimal fractions in their everyday experiences; thus, from a teaching perspective, decimal fractions are easy to teach. In such countries, students in the first few years of schooling deal with numbers up to 100 and then move into decimal fractions. In countries such as Portugal and Italy (prior to the introduction of the euro), Brazil and Thailand, where the denominations of banknotes are mostly in multiples of 1000, by the third year of schooling students are dealing with numbers in the thousands but do not work with decimal fractions; instead, common fractions are used as their experience of part number. Thus the money system of a nation can impact on the ways in which students encounter number study.

Time

Time is often the most difficult concept of measurement for students to understand. Unlike the other attributes, time cannot be measured in a perceptual way. It is an historical artefact, and hence is open to different interpretations (Poole, 1998). In most Western cultures, time is measured through mechanical means such as a clock, and seasons are measured by the calendar. In other cultures, qualitative experiences rather than a calendar may be used to identify the seasons, such as when particular forms of animal life or vegetation appear. Thus the seasons can vary in length depending on these qualitative experiences (Harris,

1990), and it is not uncommon for people to talk about 'summer being late this year' or 'summer being very short'.

Two aspects of time appear in the curriculum. The first is the passage of time, where the amount of time from one point to the other is measured as minutes, days and years. The second is a point in time that is used to identify a particular moment in time—such as Monday, January or winter. The passage of time is often difficult for students to comprehend due to their qualitatively different experiences of time—think of running for one minute against waiting for one minute.

Teaching measurement

Good teaching requires the teacher to identify common patterns in students' thinking and then move that thinking forward to the goal identified by the teacher and/or the curriculum. Often the approach is referred to as discovery learning, but whether students 'discover' anything new is dependent on two key attributes: whether the student is able to construct the desired understandings planned for the lesson; and whether the teacher is able to organise the learning environment so that the student is able to construct the intended learnings. Rather than refer to this approach as 'discovery', since the discovery has been planned by the teacher, leading the students to learn desired knowledge can be framed within Vygotskian constructs, in particular using the notions of scaffolding and zone of proximal development (ZPD). Such approaches are particularly desirable within the teaching of measurement.

Most traditional teaching of measurement has been focused on the rote memorisation of formulae that are rarely understood and thus become mere tools with which to operate. Such approaches result in students constructing misconceptions and decontextualised knowledge that have little meaning. Rather than focus on the teaching of abstract concepts—such as the gradation on the ruler or a formula for area or volume—a scaffolding approach requires the teacher to identify students' current patterns of thinking and then to create learning environments that support the development of more complex understandings.

It is important when planning a lesson or sequence of lessons (a unit) in measurement that the intended learning outcomes are clearly articulated to enable subsequent activities to be identified as linking into the learning outcomes. Being able to articulate what the teacher wants the students to learn is fundamental. Of course, students will learn more than is intended, but there is an expectation that there is some particular learning that is key to the teaching. By appreciating what is prerequisite knowledge and what comes later, teachers are able to plan to include all the students in the class. Hence models of how students learn measurement are critical as they allow teachers to plan for students who are not grasping the concept, as well as for students who are well beyond the understandings of the majority.

A number of key considerations should underpin work in measurement:

- Students should measure frequently and often. The exercises they undertake should be based on real-life tasks rather than exercises taken from textbooks. This gives them an appreciation of why we measure and why it is important to measure.
- Students should be actively involved in measuring activities rather than passively observing. The activities should encourage discussion so that students can refine their measurements and skills of measuring. Insofar as planning is concerned, this means ensuring adequate resources are available for the students (as is having a high level of tolerance for noise!).
- Links should be made to other areas of mathematics and to other curriculum areas.

● Conceptual development model of planning in measurement

A model commonly used in planning for teaching of any measurement topic delineates five phases:

1. identifying the attribute
2. comparing and ordering
3. informal measurement using non-standard units
4. formal measurement using standard units, and
5. application to problem-solving contexts.

The basic premise of this model is the underlying complexity of measurement concepts—much like the development of number and operations. The model follows a logic from simple recognition of attributes through to formal measurements and the application of formulae in problem-solving, providing a structure for planning learning experiences for students. The model is not prescriptive or grade/age specific, but provides an outline for a general process through which the teaching of measurement can occur. While the model goes some way towards describing the learning process for many students, learning is highly complex and cannot adequately be accounted for by this simple sequence.

The model therefore should be seen only as a guide for planning; teachers also need to recognise that students do not always follow its logic, since prior learning and experiences cannot always be accounted for adequately. As students enter the classroom from a wide range of backgrounds and experiences, the latent learning that has occurred outside the school can result in some students having developed considerable intuitive understandings of measurement concepts and skills, and others still having very little understanding. Where 'anomalies' occur, teachers should be ready with alternatives for students.

Identifying the attribute

In order that students are able to undertake learning in any particular measurement strand, they need to be able to recognise the attribute that is under investigation. If one considers the teaching of length, for example, there are often other distractors in the activity, yet the students are expected to identify the length attribute over all other attributes of colour, texture, purpose, and so on. Unless the students can isolate the attribute being taught from all other attributes, it is unlikely that they will be able to understand any further developments of the concept. For example, if a number of items such as a ball, a tennis racquet and a fishing line are used when talking about length, the concept may be very difficult to grasp for a student who is considering the recreational aspects of the items, or their hardness, or their colour,

Free play experience with volume to identify the attribute

Early years experience comparing and ordering volume

or any other attribute. It would be best to use only very similar items (e.g. pencils) that differed in length. In the very early stages, it would be best to use the same item in different lengths—all red straws or the same type of pencil—in order to reduce distraction. That said, research from the Early Years Project in Victoria (Clarke et al., 2000) has shown that many students come to school with a reasonable understanding of the length concept. The same cannot be assumed for other aspects of measurement.

Comparing and ordering (using direct and indirect comparisons)

Once students have been able to identify the attribute, they need experiences to compare and order (or seriate) objects according to that attribute. Beginning with two objects, students can talk about which is taller/shorter (for length), hotter/colder (for temperature), and so on, again depending on the attribute to be compared. More items are introduced progressively, and students order according to the attribute. Comparisons initially are made directly, and then through an intermediary tool (indirect comparison). The use of indirect measures creates a need for informal measurement.

Informal measurement using non-standard units

Initially, everyday experiences form the basis for measurement—hand prints for area, cups for volume, steps for length—in order to develop principles for effective measuring skills and in preparation for the introduction of formal units. Using informal measurement tools, students have access to experiences that allow them to recognise that different units can be used for different contexts—for example, while it is effective to use paper clips to measure the length of a book, a paper clip would not be a good unit for measuring the length of the room. Such experiences prepare students for the arbitrariness of formal units of measurement.

Similarly, experiences with informal units create a need for common formal units—for example, when two groups of students measure the same item with different units, they get different results. Through showing the items are the same length but have different scores for their recorded lengths, it becomes apparent that common units of length are needed when comparing items.

The use of non-standard units must be developed over time, and there are six concepts that are foundational to measuring with non-standard units. Teachers should recognise these very different concepts and plan their lessons to ensure that a range of activities is undertaken in order that such skills are developed.

- Measuring involves both a number and a unit of measure—students must be able to count how many units and include the unit of measure—for example, five paper clips, fourteen steps, six hand spans.
- Measurements can easily be compared if the units of measure are the same—that is, items can only be compared if the same unit has been used. Young children often rely on just the number to compare items—six paper clips may be thought of as being bigger than three rulers.
- Depending on the context, one unit may be more appropriate than another. Measuring the capacity of a bucket with cupfuls of sand may be effective, but cupfuls of sand would not be a good unit of measure for the capacity of the sandpit.
- The size of the unit of measure is inversely proportional to the number of units. When using small units to measure an item, the count will be higher than when using large units for measuring the same item.
- Communication of measurement is most effective when standard units of measure are used. Non-standard units of measure do not communicate measurement effectively—books, paper clips, blocks, steps, cupfuls, and so on. can vary in size so communication is not effective. Using standard units of measure, such as centimetres, grams, millilitres, and so on, allows all parties to appreciate the magnitude of the objects being measured.
- More exact measurements can be undertaken with smaller units—most measurements are estimations: 'about 6 cm', 'between 5 and

6 kg', and so on. When smaller units are used, greater accuracy can occur—'under a metre' is not as accurate as 'between 60 and 70 cm'.

Formal measurement: Using standard units of measure

Having created a need for standard units of measure and developed principles for measuring, students are introduced to standard units of measure for the various attributes. In concert with the standard units being used, students need to develop the appropriate language for such measurement, along with the correct symbolic representation for units of measure (such as cm^2). Estimation of quantities is also central.

The experiences that the teacher organises for the students should ensure that they develop a sense of how big the units of measure are—for example, how heavy a kilogram is, how much a litre is, how long a metre is, and so on. These skills are important for estimation. Similarly, there are key referents that students should also develop. By knowing what a 2 litre container of milk looks like, a student is able to estimate the volume of another item by comparing it with this item. Similar referents occur for height: 'I am 1.5 metres tall and you are a bit shorter than me, so I think you are about 1.45 metres'; 'My dog weighs about 15 kg and that dog is about the same, so it is probably about 15 kg too.'

Application to problem-solving contexts

Once students have developed and are confident with formal units of measure, they can be introduced to applications of the units to enable the development of formulae. In subsequent years, particularly in secondary schools, the development of formulae with specified values gives way to the use of variables. This stage should not be seen as entirely separate from the others, as if problem-solving should only be undertaken once the concept is fully and formally developed. Many of the earlier stages should include problem-solving in their planning. This stage tends to focus very heavily on problem-solving in terms of teaching methods.

The value of this model of concept development is that it progresses through the primary school years. It allows teachers to develop an understanding of what should have happened in earlier years and what is to come later. By having a deeper understanding of where things come from and where they are going, teachers can focus on their role in developing measuring concepts and skills. For example, it is possible to think that teaching area is about the formula 'length $\times$ width', where this is the memory of the teacher's experience. However, using the model just discussed, considerable work is undertaken before the formula is introduced. Indeed, in a model such as this the formula would likely not be taught; rather, the sequence of activities leading up to introducing the formula should mean that the students 'discover' it for themselves. Through discovering or inventing it, the formula becomes personally meaningful to students, and hence is more likely to be retained (and less likely to be confused with the formula for perimeter). Through a process of progressive scaffolding, the learning experiences provided by the teacher create an environment where the formula appears as a 'natural' development from the experiences provided.

Limitations of the concept development model

As with any model that is being used in the complex processes of teaching and learning, teachers should have some sense of its limitations. Such critique is necessary, for while the model serves a purpose it often reflects the ethos of various stakeholders. By seeing the model as unproblematic, it is at risk of becoming dogma and embraced without the recognition that in some cases it may not work. In such cases, it can put students at risk of failing and hence the teacher at risk of criticism for not understanding the intricacies of the learning processes of some students.

This model has as its root a notion that learning occurs in a sequential manner, suggesting a linearity in development that may hold true for a significant number of students. However, learning is a highly complex process, one that is often far from linear, and the linear model may deny the complexity of thinking and development in students whose experiences may be very different from those of the other students in the class. Imagine a student who has spent considerable time with his parent, who is a tiler. This student's spatial awareness, tessellation knowledge and concept of area could be expected to be significantly higher than that of other students in the same class but without the formal understandings expected in such a sequence.

To think of the sequential process of this model as 'natural' is highly problematic. While teachers may confirm that this is the learning sequence they observe in their classrooms, it may not necessarily represent the 'natural' learning process. It could be that the progression that is noted is a reflection of teachers' experiences in schools where measurement is concerned. If teachers are organising learning experiences in line with this sequence, it makes sense that observation of student behaviour will reflect their experiences. If the scaffolding provided by the teacher supports particular outcomes, then it is hardly a natural process. Providing experiences with blocks where students have to count them in order to see how long an item is, and then providing a situation in which their counting is not effective—such as some students using the blocks to count the length of a book while others use paper clips—so that the students recognise the need for a common unit to measure and compare with, is not to move naturally into the next phase of the model. Rather, the teacher has contrived a situation (either intentionally or unintentionally) whereby the students have come to recognise the need for the common units.

The five stages noted in this model can be condensed, as there are overlaps between the various levels. Rather than seeing them as discrete skills, they should be viewed as existing along a continuum. For example, identifying an attribute is often undertaken in concert with comparing the attribute against others in order to identify it. To identify mass is difficult without comparing one object against another—which is heavier, which is lighter—while keeping the focus on mass rather than volume or length. Similarly, the introduction to informal units links closely with indirect comparison. Students may decide to see 'how many rulers long' each of two items is as this is the most convenient item for them. That is, the process of indirect comparison has led to the use of informal or non-standard units of measure. The same process can occur as the students move into formal units of measure—when they are measuring how many rulers long the desk is, a need for part measures becomes apparent, so they note the desk is three rulers and 22 cm long, indicating that the table ended at the 22 cm mark.

Despite the limitations of the process, and the fact that the model is more useful as a planning tool rather than a reflection of learning, the concept development model is helpful in planning learning experiences for students.

As length is the most commonly taught measurement attribute, considerable space is spent on this topic to illustrate in detail how the concept is developed through the five stages. Subsequent sub-strands are dealt with in less detail.

Length

Length is the most easily perceived attribute, and most children come to school with some understanding of length and its associated language.

- Early experiences with length should focus on the language of length, including classifying objects according to whether or not they are short or long, thick or thin.
- Two key skills need to be developed: direct comparison, where items are compared directly against each other (e.g. placing two pencils or children side by side) to see which one is longer/taller or shorter; and indirect comparison, where an intermediary device is used. Indirect comparison might consist of using a body part—such as the door is ten hand-spans long and the desk is five hand-spans long so the door is longer than the desk. Indirect comparison is needed when measurement of two objects against each other is not possible. It is also the precursor to developing the need for intermediary devices, thus eventually leading to formal tools for measuring.
- Comparing and ordering lengths can be undertaken through the following sequence:
  1. Initially ask students to copy a sequence that has been constructed.
  2. Ask the student to place an item correctly at one of the ends of the sequence.
  3. Ask the student to place an item correctly in the given sequence (somewhere other than an end).
  4. Order three or more items. Once students can consistently order three items, more items can be introduced progressively.
- Students need to be able to make sure that informal units are lined up in a straight line, that there are no gaps between units, that the starting point is level with the item being measured and that counting of the items is accurate. These comprise a new set of skills to be learnt. Blocks that link together, such as Unifix or interlocking blocks, are very useful for the early experiences with non-standard units, as they hold together in a relatively stable manner. Teachers

should ensure that there are sufficient blocks for all students when undertaking these activities.

- Students place a number of units along the item to be measured and then take one of the earlier units and place it in the line, counting as they go (iteration). Some students will prefer to use marks to keep count of their work. This is a complex skill and should not be introduced too early; if students are having difficulty, leave it until a later stage. It is a skill that is useful later on when they need to measure lengths that are longer than the instrument they are using, such as a ruler.

- The history of length can be implemented—the cubit, the foot, a span, a digit.

- Students should develop a sense of the size of a centimetre and metre. A commonly used referent for primary school students is the student's and other significant people's height/s.

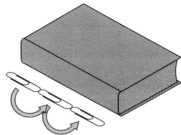

Using paper clips to measure the length of a book

- A kilometre is more difficult to gain a sense of as it is such a big quantity and can be experienced in qualitatively different ways—for example, driving to school is quite different from running the same distance.

Student measuring length using a trundle wheel

- A common mistake students make when using an instrument such as a ruler is not lining up the item to be measured with the 0 point on the ruler. They may either use the start of the ruler (the section that precedes 0) or line the item up at the 1 cm mark. When introducing standard units, it is best to use a ruler that has only centimetre gradations on it (rather than ones with millimetre marks). This allows the student to focus on a single unit of measure—a centimetre. Once they are competent with this instrument, they can use rulers with millimetre divisions.

- Having developed effective measurement principles, problem contexts can now be developed. Often this involves measuring and calculating perimeter (the length around an object). A difficulty that often arises with the teaching of perimeter is that students can become very confused between area and perimeter.

Area

Teachers need to plan for activities that allow students to identify area as being the amount of covering over a surface, along with the appropriate language of area. Activities could include covering flat objects such as tables, mats, books, and so forth with paper, tiles, base 10 block (MAB) flats or other objects. Pasting activities using standard-sized pieces of paper (such as confetti for small objects) give students a sense of which is bigger (i.e. they need several pieces of confetti to cover the one original shape). Leaving no gaps when covering areas with shapes is also very important to promote conceptual understanding of area measurements, and provides the foundation for understanding of standard units for area.

Being able to conserve area can be an important consideration, yet everyday language and practice create different learning situations. Teachers often reorganise the furniture in their classrooms to give them more space, but this does not create more area, just a better use of the available space.

T E A C H I N G   I D E A

Take an A4 sheet of paper and cut diagonally. Reorganise the shape so that it looks like the second shape here.

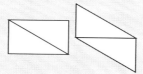

Which is bigger in area?
Justify your answer.

● Teaching strategies

- Teaching area using tessellation (which links with the space strand) is useful, particularly when regular shapes are used (rectangles, squares), as this allows the student to build rows of blocks, leading towards the development of the formula. This process links in with number study, as students need to have well-developed counting skills in order to move towards developing an intuitive understanding of the formula. Skip counting and multiplication facts are almost prerequisite knowledge.

- Grid paper can be a useful tool for implementing formal units of measure. Centimetre-square paper can be used for drawing or tracing shapes and then counting the area. The grid paper can also be copied on to transparencies so that students can overlay it on shapes and count the number of squares. It is useful to begin with whole numbers and progressively build to more complex numbers and shapes (including circles).

- Use large sheets of newspaper to make square metres so that larger areas can be covered.

- Create situations where students can 'discover' formulae for various shapes. Allowing them to discover a formula will make it more meaningful than if it has been rote learnt.

- Large areas, such as hectares and large land/sea areas, are difficult to conceptualise. World mapping exercises can be fun and illuminating for students. Compare the area of Australia with that of the United States and the United Kingdom. Students often think that countries they hear a lot about must be big in terms of area. It is often very surprising to learn that Australia and Europe are almost the same size; likewise, the United States and Australia are similar areas. Compare how many times the United Kingdom fits into countries such as New Zealand, Canada, Australia or Kenya. These activities have application to social studies.

T E A C H I N G   I D E A

Using Tasmania as a unit of one, estimate how many times it will fit into each of the other states.

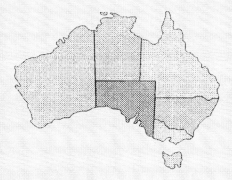

Pair and share, refine results.
Provide the areas of each state.
Using a calculator, check the estimate.

Volume and capacity

The learning activities associated with volume and capacity focus on the measurement related to three-dimensional shapes—properties of solid shapes or the amount of space an object takes up, or the amount a container can hold.

● Teaching strategies

- Begin with lots of activities involving packing and filling so as to develop a sense of volume and capacity. Fill containers with water (capacity), then immerse other objects to see displacement (Archimedes' Principle) (volume).

- In order to develop comparisons, use nesting boxes, babushka dolls and other containers that fit inside each other for direct comparisons of volume. Use items such as rice to fill one container, then pour the rice into another container and see whether there is leftover rice or leftover space to identify bigger or smaller items for capacity.
- Displacement of water can be done—items of greater volume displace more water than items with smaller volume.

T E A C H I N G   I D E A

Take an A4 sheet of paper and make the biggest container you can from it. Keep the task open ended so that a range of shapes can be made (boxes, cylinders, etc.).

The results can create considerable discussion on what 'container' means—for example, does it need to have ends?

- Informal units of measure can include matchboxes fitted into other boxes (volume), or cups, jugs and spoons for liquid measures (capacity). Get the students to record their findings.
- Using cubic centimetre blocks (base 10 blocks) leads into formal measuring units as the unit/one block is 1 cm³. Use the longs, flats or large cubes of the base 10 blocks to fill items.
  - Benchmarks and referents should be developed through linking to common everyday objects— 1 litre of milk; a 4 litre container of ice-cream; a 375 mL can of soft drink; a cubic centimetre. Once the litre has been introduced, fractions of the litre are introduced—500 mL, 250 mL, and so on.
  - Larger units of measure can be difficult but council or utility company account notices can be useful as they report water usage in kilolitres.
  - Organise activities so that students can discover the formula for volume (area of base × height) as it links to area (base of the shape). Centimetre cubes can be useful for this. A variety of solids should also be explored.

How many people fit into a cubic metre?

- Conversion between units of measure should be undertaken. Rather than relying on rote learning, provide experiences so that students can identify the relationships between length, area and volume and how these are intertwined. Ask students to determine how many cubic centimetres there are in a cubic metre: use a cubic metre and place base 10 (MAB) blocks along the bottom (10); they can then see that for the base, there are 10×10. This is then multiplied by another 10 as the cube will be 10 deep, making for a total of 1000 MAB blocks. As there are 1000 cubic centimetres in one block, calculate how many there are in a cubic metre. How many millilitres are there in a cubic metre? Linking the visual, kinaesthetic and logical learning approaches enables students to make better sense of the relationships and numbers.

Mass

When students begin study in this area of the curriculum, the terms 'mass' and 'weight' tend to be used interchangeably, but as they progress through the primary school years it is important that they come to recognise the difference between the two concepts (as discussed earlier in this chapter). This helps to prepare them for the more specific work of secondary science experiences.

Young student weighing using 'beam balances'

● Teaching strategies

- Free play with objects of varying masses.
  Activities can involve pushing, hefting (lifting) and balancing. Playground equipment is useful as it enables young students to link mass with the workings of the seesaw.
- Use tools (such as balances) or skills (such as hefting) to identify heavy and light objects. Begin by using objects where the masses are quite different in order to sort and order mass. As students become adept at identifying large differences in mass, move to smaller differences.

- Make your own balances with buckets, coat hangers and the like and use them to identify different masses by comparison.
- When using equipment such as balances, use informal units such as blocks, pencils, marbles, etc. to measure the mass of objects. Students develop the skills for using balances so that they recognise that when the balance is even, the masses are the same. Using informal units allows for indirect comparisons. Encourage the students to record their measurements.
- The language of mass is common—most students will have familiarity with kilograms and grams—so that their introduction to formal units follows the other sub-strands. Students need to gain experiences with recognising the mass of 1 kg and activities need to be included to develop referent points—such as 'I weigh 22 kg'. Weights of 1 g are difficult but other multiples are useful, such as 30 g, 100 g, 500 g and so on. Provide lots of experiences with weighing everyday items. Conversions between units of measure should be integrated into the work being undertaken.
- Explore the relationships between volume and mass. Explore packaging of goods and the transportation effect on foods.

Time

Two aspects of time need to be considered—a point in time (day, time, event) and the passage of time. These learning experiences require different teaching approaches. Time is often the most difficult attribute for students to identify in a quantitative sense, since the qualitative experiences are so different. Waiting for a minute while your tooth is getting filled is very different from a minute at a birthday party. The everyday language used in families and other social contexts also results in confusion—for example, a parent who is talking to a friend at the supermarket and tells the child that they will only be a minute.

For the passage of time, activities include counting or clapping while an event occurs. When there are considerable differences in time elapsed, young students can identify this, but the closer events are in duration, the less efficient time estimation is. For a point in time, events such as birthdays are useful for months, and focusing on the day in an informal way helps to draw attention to the day of the week—'What do we do today (sport, music)? So what day is today?' This strand often poses a point of confusion for some students. The language of duration of time links with the length sub-strand and can be difficult to understand: how long is it from home to school—3 km or 10 minutes? How long did it take to cook dinner? Common social remarks can provide rich examples of this aspect of time: how far is it from Sydney to London?—a medium-sized book.

As time is an integral component of contemporary life, there are considerable opportunities for teachers to teach this aspect of the mathematics curriculum in an incidental manner so that it is constantly reinforced. What is the time now? How long before lunch? We need to work on tidying up as we only have five minutes before play. Accurate references to time should be made, thus avoiding confusing everyday nomenclature of 'we only have a couple of minutes left'.

● Teaching strategies

Passage of time

Early experiences focus on informal units of measure, such as clapping or using egg-timers. Events with common starting points: 'Who took longer to win the race?' (i.e. direct comparison) are introduced first and followed by indirect comparisons, where the duration of an event is measured and compared. Informal units ('How many more sleeps to my birthday?') are used to create the need for formal units of measure (seconds, minutes, hours, days, weeks, etc.).

Once formal units of time have been intro-duced, students can measure the passage of time using clocks and stopwatches. Comparisons can be made to see which events take longer or shorter periods of time. At some stage—usually towards the upper years of schooling, once number and operations have been mastered—students begin calculating with time. This can pose problems since there is often a tendency to work in base 10

Student-constructed candle clocks

rather than in the various units of time. When introducing this aspect of time, teaching episodes should be carefully planned, much like teaching operations. Begin by using whole units of time so that there is no need for trading (e.g. How many hours between 4 o'clock and 8 o'clock?) and once this has been mastered, move to trading (e.g. if it is 4.30 p.m. and the train leaves at 6.10 p.m., how long do I have to wait?).

Point in time

- Students' experiences with points in time are developed in progressively smaller units. 'Morning', 'afternoon' and 'night' precede more specific moments in time. Students begin to move to times of the day that relate to their everyday lives. In the past, there has been an expectation of what would be appropriate times of the day for different events to occur, but increasingly teachers recognise that their students' lives can be substantially different from their expectations. Requiring minimal reading skills, pictures can be used to depict daily events for students to place in order (having breakfast, going to school, having dinner, having a bath, going to bed).
- Birthday graphs are a useful teaching tool for both time and data sub-strands.
- Use calendars, and relate these to longer periods of time (such as the seasons). Decades and centuries must also be learnt, but these are often difficult for younger students, since when 'decade' is introduced, many of the students are not even that old!
- Long epochs of time are introduced, allowing the introduction of concepts of AD and BC.

Time as linear or cyclical

In Western thinking, time is considered a linear process. For many young students, this results in thinking that if their birthday is in, say, June they have to wait longer for their birthday to arrive than their friend whose birthday is in March. In part this is because the calendar is presented as a document from January to December.

● Time-telling

- This skill takes considerable time for students to develop. It links closely with number knowledge (fives multiplication facts, fraction knowledge), so these concepts need to be developed in order to make most gains in time-telling.
- Early experiences with clocks—analogue and digital—focus on features of clocks, which builds into time-telling of o'clocks and half-pasts. This builds into quarter-turns before moving to minutes past and minutes to.
- Linking digital and analogue times is an important part of time-telling. Students' early experiences with digital time see them saying times such as 'fifty minutes past three'. Bingo games can be useful for this linkage, including the use of clock faces (analogue and digital) as well as written times in both numerals and words. Developing notions of a.m. and p.m. precedes learning 24-hour time. Once students are familiar with these aspects of time, international times can be introduced. Investigate time lines—Greenwich Mean Time, International Date Line, making international phone calls.

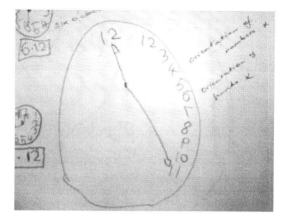

Student's early drawing of a clock

- Practical applications of timetables (train, bus, ferry) that are relevant to the students' experiences are introduced. These experiences build towards calculating how long trips take.

T E A C H I N G   I D E A

Time bingo

Create a series of cards displaying different times in different formats. Play as bingo.

3:00	(clock showing 3:00)
3 o'clock	Go home from school

Temperature

There are a limited number of experiences that can be provided for students to identify this attribute. These usually consist of using sensory experiences of touch to feel whether things such as water, the asphalt on the netball courts, and so on are hot or cold, as opposed to other attributes such as prickly, soft, fluffy and so on. Students have a relative sense of temperature, so development of language is important in the early years—hot, cold, warm, freezing. This would be done before introducing thermometers.

● Teaching strategies

- Provide practical experiences with feeling temperatures—cold, warm, hot—and ordering of the same temperatures.
- Use cut-outs of different climates (North Pole, deserts, tropical areas) or the seasons (summer, winter, autumn, spring) to order.
- Keep a daily weather chart and plot the temperature—either measured by the class or from the daily newspaper.
- Use thermometers to measure temperatures.

Money

The money sub-strand is potentially difficult for some students when discussing value. What young students value can be very different from what older students value. Value and worth can be two mutually exclusive constructs. Consider a home by the ocean—for someone who loves the sea, the house is worth more than it would be to someone who loves the outback. Unlike other areas of the mathematics curriculum, the concept of value is relative rather than absolute. For students, what is central is money-handling skills; thus this topic is more commonly referred to as money than value. It is interesting to look at the teaching of number in another country and linking it to the monetary system in that country. For example, in Portugal and Italy (pre-euro), the currencies had very high numerical denominations (which can be conceptualised as being akin to Australian currency if money was in cents only rather than dollars and cents). The units of values (escudos and lire respectively) were very high (e.g. shoes could cost 3000 lire), so the decimal part of the currency was not in circulation. As such, decimal numbers did not feature in these countries' study of number. The introduction of the euro offered unique opportunities for teachers and researchers to study the transition and development of thinking in number.

Early money activities can be difficult for Australian students since the coins are not proportional. While the silver coins increase in value as they become bigger in area, the gold coins reduce in area as they increase in value. In contrast, New Zealand's gold coins increase in size as they increase in value.

The money sub-strand is linked strongly with the study of number. It has good application of decimals since dollars can be considered whole units, and the cents as parts of the whole.

- Discuss items that have value to the students—such as trading cards (Pokemon cards, football cards) and how some cards are worth more than others; discuss items that have high or low value (cars and sweets).
- Create a class shop and have students put prices on items. Use money (plastic, paper copies) so that students can develop coin-recognition skills.
- Values and equivalence feature in the work undertaken so that students will be expected to know the value of particular coins and notes, but also be able to recognise equivalence in values—that two 50 cent coins are the equivalent of one dollar. Bingo games are a useful tool for teaching this aspect of number.
- Operations with money link in with number work. Initially, work will be in just one unit of measure (cents or dollars) until the students have become fluent with decimal fractions. This knowledge can then be applied to mixed units of measure. As number work develops, so too can money study. For example, as students work with multiplication of mixed numbers, they can do the same operations with money.

Making connections between sub-strands

● Area and perimeter

Students often get confused when area and perimeter ideas are put together (Gough, 1999). In part, this can arise from poor understanding of area and length. Activities need to be given in which students' misconceptions can be challenged. The use of a geoboard can help students to see that the perimeter may be held constant (with a piece of string of a particular length) but the area can change depending on the shape—24 cm of string can make a shape of 6 × 6 cm; 3 × 9 cm; 4 × 8 cm.

Using grid paper, students can keep area constant using blocks, and then find out the different perimeters of the shapes they construct. In 'Teaching idea—Grid shapes', the area is four units and depending on how they are placed, a range of different perimeters can be found.

T E A C H I N G   I D E A

Grid shapes

Using a sheet of grid paper, make as many different shapes as possible with four blocks. Colour the shapes different colours. What is the area and perimeter of each of the shapes? Record the results in a table. What do you notice?

● Volume and surface area

Just as students get confused with perimeter and length, many students also become confused with volume and area. Folding a piece of A4 paper into quarters in both directions can result in different volumes, as can folding A4 paper into a cylinder in both directions. Students can apply this knowledge to packaging and environmental issues so that they can find out how best to pack items with minimum surface area and maximum volume. This knowledge is also useful for understanding how food manufacturers can increase the price of goods without the public becoming aware of it. Current practices have seen a number of manufacturers reducing the content of containers but changing the shape (surface area) so that they look the same, and charging the same amount of money.

TEACHING IDEA

Volume and shape

Using the same number of blocks, create different shapes.
What is the surface area of each shape?
Record results.

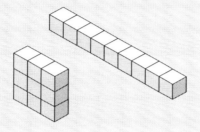

● Volume and shape

Just as with the other examples, shape and volume can develop some interesting misconceptions. This relates to the example above, but without formally acknowledging surface area. Allow students to construct various shapes using blocks. Keep the number of blocks constant (i.e. volume) and see what shapes they can make.

● Volume and capacity

The connection between liquid measures (mL) and solid measures (cm³) often escapes students. This restricts their appreciation of the rich relationships in the metric system. Providing opportunities where these relationships can be explored provides a rich basis for later work, particularly in science. It also helps students to move away from rote memorising of tables of equivalence and allows them to recognise the connections between volume and capacity. Provide a range of items where students can calculate (or count) the solid measure. Fill with water and then pour into a graduated container to see the corresponding liquid measure.

Similarly, have particular measures of water in the graduated jug and then pour into the 1000s block container to see where it corresponds. Measures of half-litres, quarter-litres, etc. will be easy for the students to see.

T E A C H I N G I D E A

Volume and capacity

Pour water from a 1 litre jug into an empty 1000s cube. This allows students to see that 1 litre (1000 mL) is the same as 1000 cm³.

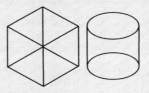

● Volume and mass

The beauty of the metric system is its cross-linking. The activities undertaken in the item above can be extended to measuring mass. Using water, have students weigh their containers (this is later subtracted unless the scales have a zeroing mechanism). Find out how much different measures of water weigh. This does not hold true for other items, where the density will alter the weight. Students can fill the 1 litre or 1000 cm³ container with different items (feathers, stones, metal, sticks) to see how much these weigh. This brings about the early concept of density, which is later explored in science and mathematics.

Teaching measurement in the middle years

In the middle years, students' understanding of measurement cannot be taken for granted. This means that teachers must continue to provide hands-on, active learning experiences to ensure students' conceptual understanding of measurement topics becomes consolidated. Assessment studies have found that students in the middle years will make errors in measuring by not aligning the zero on the ruler with the edge of the object being measured. Further, students' knowledge of formulae for measurement calculations is tenuous, as students have difficulty understanding why formulae cannot be applied in non-standard contexts (Battista, 2007). In summary, 'the research indicates that many students' measurement reasoning is superficial, with poorly understood procedures or formulas substituting for deep understanding' (Battista, 2007, p. 893). In the middle years, the teacher's role is to provide opportunities for students to see the limitations of their formulaic measurement knowledge. The following situation is a great way for students to question their knowledge of area and perimeter: 'If I have a fixed length of fencing to fence my garden, will I get the same area regardless of the shape of my garden?' Tasks such as these require students to think about how they might convince someone of their response. They need to use concrete materials, and most likely some grid paper, so that they can argue their solution. Rather than completing pages and pages of exercises where students have to find the area, the volume, the perimeter, and so on of various objects, look for activities that require them to think about the situation and their measurement knowledge. The modern classroom textbook is generally replete with such good investigative activities that are frequently overlooked. Finding and using these types of problems is the essence of teaching measurement in the middle years.

■ REVIEW QUESTIONS

13.1 List the general sequence for teaching any measurement sub-strand. Provide illustrations of the sequence.

13.2 List and explain three considerations when teaching time.

13.3 Current research indicates the importance of making connections in mathematics. How could you make connections between some of the topics in measurement?

13.4 The teaching of measurement is based on a Western world-view of quantifying. Many other cultures see the world differently. If you were working in an indigenous society that placed a greater emphasis on quality, what might this mean for teaching this aspect of mathematics?

Further reading

Aldridge, S. and White, A. (2002). What's the time Ms White? *Australian Primary Mathematics Classroom*, 7(2), 7–12.

Friederwitzer, F. and Berman, B. (1999). The language of time. *Teaching Children Mathematics*, 6(4), 254–9.

Monroe, E.E., Orme, M.P. and Erickson, L.B. (2002). Working cotton: Toward an understanding of time. *Teaching Children Mathematics,* 8(8), 475–9.

Nitabach, E. and Lehrer, R. (1996). Developing spatial sense through area measurement. *Teaching Children Mathematics*, 2(8), 473–6.

Weinberg, S. (2001). 'How big is your foot?' *Mathematics Teaching in the Middle School*, 6(8), 476–81.

Whitby, K. (2001). Teaching and learning about length. *Prime Number*, 16(4), 9–13.

References

Battista, M. (2007). The development of geometric and spatial thinking. In F. Lester (ed.), *Second handbook of research on mathematics teaching and learning* (pp. 843–908). Charlotte, NC: Information Age.

Clarke, D.M., Sullivan, P., Cheeseman, J., and Clarke, B.A., (2000). The Early Numeracy Research Project: Developing a framework for describing early numeracy learning. In J. Bana and A. Chapman (eds), *Mathematics Education Beyond 2000* (pp. 180–7) Fremantle, WA: MERGA.

Gough, J. (1999). Perimeter versus area: Fixing false assumptions. *Prime Number*, 14(3), 19–22.

Harris, P. (1990). *Mathematics in a cultural context: Aboriginal perspectives on space, time and money.* Geelong: Deakin University Press.

Moore, D.A. (1999). Some like it hot: Promoting measurement and graphical thinking by using temperature. *Teaching Children Mathematics*, 5(9), 538–43.

National Council of Teachers of Mathematics (1989). *Curriculum and evaluation standards for school mathematics.* Reston, VA: NCTM.

——(2000). *Principles and standards for school mathematics.* Reston, VA: NCTM.

Poole, R. (1998). Calendars, calibration and culture. *Mathematics in School*, 27(5), 14–16.

CHAPTER 14

Chance and data

What are chance and data?

Probability and statistics are encountered with regularity in everyday life. It is generally accepted that this is one of the key numeracies in the new millennium because of the data-saturated society in which we live, the rapid changes occurring in society and work, and the growing use of computer technology. Many educators and social commentators argue that the significant changes occurring in Western societies are akin to the changes that resulted from the move from agrarian communities to the organised workplaces of the Industrial Revolution. Contemporary theorists and educationalists argue that students need to enter this world able to make sense of the multifarious forms of information presented to them. For mathematics teachers, the task becomes one of preparing students for a society in which they need to be able to make informed decisions about whether or not a particular medical procedure or treatment is worth the risk; their money and what to invest in; whether to make purchases on any number of credit options or delay purchase; assess risks associated with various types of purchases with cars or other items, and so on. To do this, they need to be able to evaluate the likelihood of events occurring (probability) and use information (data) to make the most appropriate decisions. As much of the information they seek will come from the World Wide Web, internet-related activities should be included in teaching (see Dixon and Falba, 1997 for some examples).

● Statistical literacy

A renewed focus on statistics and probability in the new Australian curriculum underscores how an understanding of chance and data is the linchpin of effective survival in the world beyond school. It is therefore vital that key skills and attitudes be developed. The emphasis of this strand goes beyond exploring probability and constructing graphs. It is far broader, and directly targets the development of statistical literacy. This form of literacy is central to the life skills with which students should exit from school. The characteristics that define this literacy include being able to identify the types of data that need to be collected; undertaking and subsequently organising the data collected; representing the data in ways that make sense and result in high levels of readability; and being able to interpret and critique data. Everyday literacy and numeracy are combined in this strand—today, even reading the newspaper involves interpretation of considerable statistical numeracy. Being able to interpret *and* evaluate the information being presented is a key life skill.

One of the more common experiences students will have of mathematics in their everyday lives is the interpretation of data—through the use of either 'averages' or graphs. Being able to read, interpret and analyse texts is a key literacy skill. Similarly, being able to undertake the same tasks when the texts involve mathematics is a key skill that extends the literacy demands to include aspects of mathematics. This form of literacy frequently involves mathematics related to the chance and data strand. Statistical literacy involves the elements of text reading identified by Luke and Freebody (2001), as well as a numeracy element. Being 'literate' in these contexts demands that students have an understanding of, for example, how measures of central tendency are used in everyday contexts—the 'average price of houses has increased by $40 000'; 'the average salary of workers is $350 a week'; 'the average age of teachers is 48 years'. Knowing what these figures mean is an important component of being literate in today's data-drenched world.

Identifying the writer's intentions through analysing why one measure of central tendency (or graph) was used over others helps the reader identify the purpose of the text and any tools the writer was using in order to convey particular meanings to the reader. Whereas once the data strand focused on the construction of graphs or measures of central tendency, it must now incorporate reading, interpreting and critically analysing the information (or texts) in order to deconstruct meaning and any assumptions being built into the texts. For example, different measures of 'average' are used to convey information to the public; being aware of what measures are used is important in understanding.

Much like the other strands in the mathematics curriculum, chance and data are arbitrary divisions of content constructed in order to make sense of an expansive and expanding body of knowledge. While they are presented as discrete sub-strands, there is considerable overlap between them. For example, when measuring experimentally the probability of an event occurring (such as throwing a six on a die), the throws must be collected and organised (usually as a table). In order to collect and collate the information, a knowledge of the data sub-strand is essential.

Probability

Chance and probability refer to the same construct. Probability is concerned with measuring the chance that a particular event or outcome is likely to occur. It is measured in two quite different ways. One is experimental, where students actually undertake the process (such as throwing a die to see how many times it lands on each number). The other is theoretical, where students identify the number of possibilities in total, and the number of possible times that a preferred outcome can occur. Often chance is seen in the context of gambling options, and frequently the activities undertaken in the teaching of the concept—using cards, dice, coins—have strong connotations of gambling. However, in daily life such decisions as whether to drive a particular route, whether to undergo a medical procedure or take a particular medication, or whether to stay on a casual salary or move to a permanent salary are all based on chance. In order to make decisions that can influence the quality of life in a positive way, understanding probability is an increasingly important life skill.

Data

Other terms for the data sub-strand include 'visual representation' or 'statistics'. This sub-strand refers to the collection, representation and interpretation of data. Considerable time is spent in this sub-strand on developing students' knowledge and skills in graph construction and interpretation. However, it has been recognised widely that while students may be able to construct graphs, they may not transport this knowledge to their everyday lives and use it to interpret the graphs and other forms of data presentations with which they are confronted every day. More emphasis needs to be placed on reading, interpreting and critically appraising the information that is presented through graphs and tables.

When the data sub-strand is seen to be inclusive of statistics, it incorporates aspects of data but also refers to measures of central tendency (mean, median and mode). These measures of central tendency are another form of data representation, and students need to have this knowledge in order to interpret the information with which they are presented on a daily basis—for example, to be able to critically appraise reports in the daily paper and to assess the legitimacy of such claims as 'the average number of children per family is 1.7'; 'the average Australian child is overweight'; 'the average number of hours of television that children watch per week is 35'. To do this, understanding the way in which 'average' has been measured is critical.

Teaching chance

The model used for planning in teaching measurement can be applied in the teaching of chance, since it is a process of measuring the probability of an outcome occurring. As stressed in the measurement chapter, organising for learning using this model is simply a planning process—and while it has value, it is not meant to be the one true model for planning. It is, however, widely used in syllabus documents for organising the mathematical progression through school, and research suggests that students progress through particular and predictable patterns of chance understandings (Watson and Moritz, 1998).

● Identify the attribute

Students' everyday experiences of chance or probability can be used as the catalyst for discussion in order to identify the attribute. Using everyday terms such as likely, impossible, maybe, unlikely and impossible, it is possible to identify the construct of probability. Using everyday experiences such as whether it is going to rain, snow, be hot; whether the students will get certain presents for birthdays or Christmas; what will be in their boxes for

lunch, and so on, can be useful for talking about the likelihood of events happening. Words that are part of students' everyday vocabulary can be listed so that they are able to build up a strong language of probability. A great deal of time during the early years of school is spent on developing an informal understanding of probability through everyday experiences. Opportunity should be given to allow students to predict the possible outcome of events and to discuss their level of confidence in their prediction (such as throwing a six on a die).

● Comparing and ordering likelihood of outcomes

Using informal language of probability, students categorise events according to the likelihood of them occurring. Beginning with terms such as possible and impossible, or certain and uncertain, or unlikely and likely, events are categorised. These could be picture-sorting activities or using simple table formats. More complex categories can be constructed using terms such as 'more likely' or 'less likely'. Simple activities can be undertaken so that students can compare the likelihood of events happening when there are only two equal outcomes (e.g. tossing a coin, taking a red ball from a jar when there are equal numbers of two colours). Students should be encouraged to record their experiences

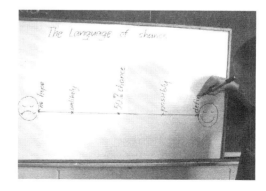

Student ordering the probability of events

on charts or tables so that they begin to see the emergence of the formal definition.

Activities can include the following:

- Toss a coin and record how many times it lands on heads and on tails. Compare which one occurred more often—heads, tails, or about the same.
- Have a bag containing balls of only two colours (e.g. red and blue). Students take a ball out, record the result and place the ball back. Repeat a number of times.
- Once two outcomes have been explored, move to experiences where there are more equal outcomes (balls of three different colours in the bag).
- Throwing a die and recording the outcomes on a table. Students often have the misconception that throwing a six, for example, is more difficult than any of the other numbers, based on their experiences of board games. Recording their outcomes can provide an opportunity for them to see that there is a relatively equal chance of each number occurring.

● Measurement of probability

Prior to the formal measurement of probability, students should have a good understanding of number study in the areas of fractions and ratio since probability is measured using these number skills and concepts. When an event is impossible, it is seen as 0 (zero) and when it is certain it is seen as 1 (one). Two processes are possible for understanding probability.

TEACHING IDEA

Toss a die 30 times. Record the outcomes in a table:

1	2	3	4	5	6

Discuss individual outcomes. Combine small group or whole class outcomes. Discuss.

Theoretical probability

The theoretical probability is tied to understanding two aspects of probability. The first component is the number of times that the desired outcome can occur. For example, when considering throwing a die, the number of preferred outcomes is considered. A single outcome, such as throwing a 4 or 1, is the usual starting activity. This can be increased to include other outcomes such as throwing an even number (2, 4 or 6) so that there are three potential outcomes, or a multiple of three (3 or 6) so that there are two potential outcomes. The second consideration is the total number of outcomes that can occur. In throwing a die, these are throwing 1, 2, 3, 4, 5 or 6 so that there are six potential outcomes in any one throw.

$$\text{Probability of an event} = \frac{\text{Number of individual outcomes}}{\text{Total number of possible outcomes}}$$

Thus the probability of throwing a 6 can be recorded symbolically as Pr (6) = ⅙. The relationship of probability fractions is seen from such symbolic notation.

T E A C H I N G   I D E A

Using two dice, what are the total number of potential outcomes? What is the probability of throwing a six? Is it the same as throwing a four? Is there a number that is more likely to come up than others? Why? Justify your answer.

Experimental probability

Actively involving students in explorations of probability provides them with the opportunity to link the components of the theoretical measures to their experiences. Throwing a die creates the opportunity for them to see the rationale for recognising the total number of possibilities, since they will need to construct some sort of table or recording tool for the numbers they throw. As they undertake the throws, the need for the table with six columns emerges, and with it the recognition of the number of possible outcomes. Using experimental situations can also prompt a more lateral thinking process. For example, earlier experiences with single die throwing can create a misconception about the probability of throwing numbers when using two dice. Often students think that there is an equal chance of throwing any number. They also think that there is a particular range of outcomes. The two data tables below show the possible outcomes when two dice are thrown. The first table indicates the actual combinations possible (36 in total) and the second table indicates the number of times each total can occur. Thus there is more chance of throwing a 7 with two dice than a 1 or 2. Using an experimental approach gives students the opportunity to challenge their existing schema based on the simpler examples given when introducing probability.

Results of addition of two dice

+	1	2	3	4	5	6
1	2	3	4	5	6	7
2	3	4	5	6	7	8
3	4	5	6	7	8	9
4	5	6	7	8	9	10
5	6	7	8	9	10	11
6	7	8	9	10	11	12

Number of outcomes for each combination

outcome	2	3	4	5	6	7	8	9	10	11	12
number	1	2	3	4	5	6	5	4	3	2	1

Some situations cannot be measured theoretically, and hence are more open to experimental probability. Such situations can also lead to rich discussions on notions of fairness. Some examples of these situations would be throwing a drawing pin and noting how often it comes up one way or the other. Similarly, the well-known question of which way up a slice of buttered bread lands when it is dropped from the kitchen bench provides a rich discussion point. In this case, it would appear that there are only two outcomes, each equally likely, yet our everyday experience suggests that it almost always lands buttered side down! Asking questions such as 'Why?' or 'Is it really the case?' stimulates discussion.

● Application to problem-solving contexts

The application of probability to problem solving can involve two issues. One is the mathematical aspects of probability, whereby students can investigate the probability of horse racing or Lotto in order to decide whether or not they would gamble. Less morally motivated applications of probability can involve Gallup polls and the lead-up to elections; weather forecasting; insurance premiums (why it costs so much for young people to have full comprehensive insurance and high excess costs); applications to familiar everyday situations—leaving work at 4.00 p.m. or 4.30 p.m. and the time it will take to get home due to peak-hour traffic.

The second aspect involves critical mathematics and the exploration of the social costs and issues of gambling on the community and individuals. Investigating why people would spend so much money on Lotto with such little potential for winning, and similarly investigating social issues that arise from gambling—such as embezzlement from family trusts or employers to supplement gambling; parents leaving young children in cars while they gamble in casinos; the breakdown of families due to gambling; and the support group Gamblers Anonymous— are all issues that can be explored (and linked to other curriculum areas). Such issues appear regularly on TV news reports and in the daily newspapers.

● Teaching points for probability

Within the teaching of probability, a number of teaching points need to be observed. Embedded in the activities outlined above are some key constructs that will be integrated into the teaching process.

Outcomes and sample space

Students need to gain an appreciation and understanding of the notion of outcomes and sample space. Outcomes refer to the number of outcomes for a particular event (e.g. an even number on a die). The sample space refers to the total range of potential outcomes. For example, when using a six-sided die, the number of outcomes is 6, and the sample space is 1, 2, 3, 4, 5, 6. If the preferred outcome is 4, as there is only one 4 on the die this would mean that the preferred outcome is 1. When throwing this die, it is then reasonable to expect that there is a probability of ⅙ and that the outcome will be a 4. This is represented in the following way:

Pr (4) = ⅙.

This can also be expressed as a decimal or percentage.

TEACHING IDEA

Fair go!

Students are asked to consider the following newspaper advertisement and then write to the Minister of Fair Trading justifying why the game should or should not be allowed at the local fair.

Tim Collins from Sideshow Alley asks all the kids in school to come and try their luck on his new game. The game is for you to throw two dice and win some money. If you throw a 1, 2, 3 or 4, you double your money. If you throw a 5, 6, 7 or 8, Tim keeps your money. If you throw a 9, 10, 11 or 12, your money goes to charity. Tim will not make any money on this game as the money he keeps on his scores is used to pay the double money students get on their scores. So come on down to the fair and double your pocket money!

Independence of events

Coming to appreciate that some events are independent of others can be difficult, since common everyday misconceptions tend to sway judgements. For example, if a couple has had four sons, there is a perception that the chance of a daughter next time must be much greater. Similarly, when tossing coins, if six heads have been thrown in a row, then it is expected that the next toss will be much more likely to be tails. When students are confronted with this type of outcome, they tend to offer excuses such as 'there are more holes [indentations] on the six dots [of the die] so it weighs less, that is why it comes to the top'. The law of averages logic is falsely applied so that judgement is influenced incorrectly. In both cases, each event is independent of the others—that is, the outcome of one event does not impact on the others.

Tossing a coin is not influenced by the throws from earlier tosses. Each toss has an equal probability of resulting in heads or tails, regardless of what has happened earlier.

In some cases, events are dependent on other outcomes (as in games such as Lotto), where the probability of the second event is influenced by the first event—for example, when a ball is removed from the pool in the first draw, so that the second draw has one less ball than the first draw. So the probability of the first ball is 145 (where there are 45 balls) and that of the second ball is 144, since there are only 44 balls left. In this case, the probability of the second event is influenced by the first event.

In order for students to gain an understanding of the independence of events, activities should be provided that allow them to explore the outcomes when sample sizes are increased. When tossing a coin, for example, the outcomes may be skewed (e.g. on ten tosses of the coin, there may be two heads and eight tails), but as the sample size increases, or class results are pooled to increase the sample size, students can see that the actual outcome comes closer to the theoretical outcome (i.e. theoretically, the outcome for tossing a coin is 50/50).

Randomness

Randomness is a key aspect of probability. For an event to be random, it cannot be influenced by other factors. The notion of randomness brings in considerations of fairness. If, at the end of the term, students were to be allocated a chocolate bar, discussion as to the event being random and fair could be developed. Would the students prefer to have everyone's name in the bag? Should the teacher be able to see the name before she draws it out? Should the girls be allowed to have their names in twice? By recognising biases in the sampling process, the notion of randomness becomes apparent.

Data collection, representation and interpretation

Current trends in teaching data are informed by the wider research community, where it is noted that while students may be strong on constructing graphs, they do not appear to be as strong in interpreting from the graphs and being able to transfer this information into decision-making. The emphasis in this sub-strand needs to be on a range of ways of collecting and representing data, along with interpreting data. As with the other aspects of teaching mathematics, the role of the teacher is to guide learning in ways that are supported so that students are able to achieve the desired outcomes. Scaffolding learning is a central part of curriculum planning and organising. Teachers need to be aware of what knowledge is reasonable to expect from the students and what knowledge students will be expected to construct, and where this will ultimately lead.

● Collecting data

In order to collect data, a key decision has to be made regarding the purpose of collecting them. By knowing why the data should be collected, decisions can be made about the types of data that are most

appropriate. If data were required about the modes of transport to school, information on the types of cars the teachers drive is not the right data to collect; rather, the various forms of transport need to be identified—cars, bikes, buses, walking, scooters, etc. The sample from which the data are collected also needs to be considered—is it just the class or the school or the community? Collecting data for data's sake is a process that has been used in schools for some time, but the purpose of this needs to be questioned.

Students collecting traffic data

Data should be collected for realistic purposes— why do we want to know about how we come to school (or work) other than to make a graph? What is now central is to consider the *purpose* of the graph—in this case, it may be part of a larger project on environmental issues where modes of transport are a key issue, and hence the collection of data has a real purpose. In the early years, the purpose may link into the health curriculum, where notions of healthy eating are being considered, so collecting data on diets and lunches serves a real purpose. By ensuring the data collection (and subsequent representation) has real purpose, the activity is seen as a valuable, authentic and purposeful activity rather than simply a time-filling exercise.

● Organising data: Teaching graphs

Rather than seeing graph work as a formal process of x-axis and y-axis, contemporary approaches adopt a scaffolding approach where the need for various components emerges from problematising the process. When students are able to recognise the needs for the various components of a graph, students are less likely to see them as something demanded by the teacher but as key and relevant aspects of the graph. Prior to constructing graphs, students should be given considerable experience in reading and interpreting them. The graphs with which students work—both interpreting and constructing—should be relevant to experiences that may be immediate as well as cross-curricular (Urso, 1999). Much of the work undertaken in number study is reinforced in this strand. For example, number study of large numbers (millions) will precede work with population graphs.

Simple picture graphs

Experiences with graphs in the early years of schooling are strongly linked to the experiences of the students. Often they relate to when their birthdays appear in the year, the types of pets they have, how many siblings are in their families, what they ate at a particular meal and so on. The concreteness of these examples allows very concrete representations to be developed within the class. There is no real need for labels at this point, since the data and numbers are obvious. However, once students have become familiar with organising data in this format, the task for the teacher is to progressively delete concrete evidence, thereby creating the need for labelling. The need for a common starting line and even spacing between the columns is central to these early experiences. For the teacher, when beginning work with pictographs, it is recommended that squares be used for the pictures so that the orientation of the picture will not matter. Square Post-it® Notes are a valuable resource for this activity, as graphs can be made very easily—students draw on the data (their favourite ice-cream, what they had for lunch, what they do on

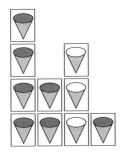

holidays) and they can easily be stuck to a baseboard. This allows for constructing many graphs quickly.

It is advisable that early work with graphs focuses on two-column graphs, progressively building to more columns as students demonstrate an understanding of the principles being developed.

Replace pictures with squares

In creating a need for labelling the x-axis, the teacher changes the pictures so that they do not visually cue the reader into seeing them as hair colour, for example. This could mean using brown, black, red and yellow pieces of paper. Other resources, such as blocks or counters, can also be used. On seeing such a representation, the reader would be unaware of what the graph was representing, thus developing the need for the axis to be labelled.

Similarly, as the columns become larger and less obvious in number, the need for labelling the vertical axis (or y-axis) emerges. In the graph shown in the diagram, students have moved on from using picture graphs, but it is unclear just what the graph is representing. In order for it to be known to the reader, the axis must now be labelled. As this is the object that is going to be measured, mathematically it is referred to as the independent variable. This might be hair colour, or months of the year, or foods eaten. Similarly, the other axis is used to represent the dependent variable as this axis 'depends' on what is the object of the count. For example, the number of students represented in a particular column is dependent on what the focus of that count is—the number of students with brown or blonde hair, or who had pies or hotdogs for lunch. As with many mathematical ideas, it is important for teachers to know the mathematics but not to teach it explicitly to students. By knowing, and hence teaching it correctly, the teacher will ensure that students will have less chance of creating erroneous knowledge.

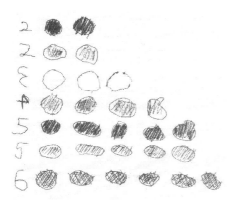

Early representations of graphing

Transfer to grid paper

At some point in time, the students need to record their graphs on paper. One of the easiest ways to achieve this is through the use of grid paper, which allows the students to simply colour in the appropriate number of squares to represent the data. Initially this might be done with different colours to represent different groups of data. After some time and experience, students will use the one colour for all columns, thus creating a need to label the individual columns.

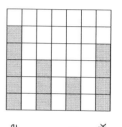

Grid representation of data

A need for labelling along the vertical or y-axis emerges as the data numbers increase. It is easy to recognise small groups but as the data sets increase, a scale is needed to enable easy comparisons. In the early stages of graphing, the scale should be a one-to-one relationship. Once this relationship has been grasped, the scale can be increased so that the students can have one square representing two, five or ten people, or whatever scaling is appropriate for the data being represented.

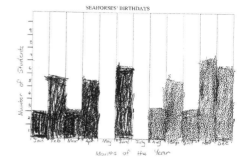

Student work sample: class birthday graph

Horizontal and vertical bar graphs should be used so students become familiar with both formats. The process of labelling the graphs and the axes is the focus of teaching, along with labelling any scales that might be used.

Construct column graphs

The final phase in the construction of column graphs is the removal of grids and the drawing of columns that correspond with given amounts on the y-axis (when using vertical columns). Simple scales can be introduced. Students clearly label the y-axis and mark in scales and numbers on it. By progressively developing the need for the various components of graphs, students can recognise their purpose, and in so doing appreciate their appearance and role. When this understanding is developed, they are more likely to include the appropriate information and modes of construction when developing graphs rather than to memorise abstract features and hope that what is included is correct.

● Types of graph

A common student misconception in the graphing sub-strand concerns the ways in which data can be displayed using column graphs. There is a range of column graphs, and their different purposes are the major reason for choosing a particular construction. One important aspect students need to recognise is whether or not bars should touch. This is often poorly understood in the primary school years, creating difficulties for students when they begin their formal study of statistics in high school. The difficulty lies in recognising the difference between continuous and discrete data, as these data types influence the type of graph constructed—histographs for continuous data and bar graphs for discrete data. Other types of graphs with which students should become familiar are noted below.

Comparative bar graphs

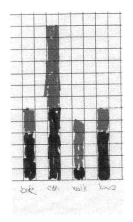

End-on-end
comparative bar graph

Comparative bar graphs are used when there is a need to compare two sets of data. For ease of comparison, the two sets of data are placed on the same graph. For example, a teacher might want to compare how a group of students has performed on one assessment piece compared with another assessment piece. Students may want to compare the rainfall for each month over two years; how many pizzas a cohort of students may have eaten over two different time periods; the pets owned by two cohorts of students.

Using the same data set, two different forms of comparative bar graph can be used. The data sets in the graphs pictured refer to students' end-on-end comparative bar graph preferences for four modes of transport to school. In the side-by-side comparative bar graph, columns are side by side so that comparisons can easily be made between the gender and the transport. The end-on-end graph allows the total transport modes across both genders to be compared.

It is critical when using these forms of graph that the data are of a form that can be compared. For example, it would not be possible to compare data that discuss income and expenditure, since these are not things that can be 'added'.

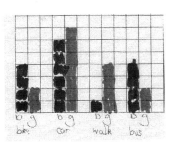

Side-by-side comparative bar
graph

Grouped frequency histographs

Continuous data are often 'clustered' so that small ranges can be used to organise them. This happens implicitly with all continuous data, although this is often not recognised or made explicit to those collecting the data. For example, when constructing an age graph, the data are collected on the age of the student so that while the 'year' is seen as the category, it is how the category is constructed that determines how the data are grouped. Consider the ages of students in a multi-age Year 4/5/6 classroom. The expected age range is probably between seven and twelve years. If this age range is broken up into months, so that the ages are more accurate, it will run from 84 months to 144 months. As a continuous range would not be used, the data are grouped and categories constructed. As discussed later in the section of this chapter on discrete and continuous data, how these data are grouped depends on where cut-off points are made. In the data table, these are up to any given birthday. Once the student has had their birthday, they move into the next category. It could have been just as easy to have made categories that ranged from halfway between one birthday and the next—that is, 6½ to 7½ years.

The compiled data can be grouped into clusters of data that are useful for representing the final outcomes. The data can then be transferred from a table format into a graphical form. The data can be entered into a spreadsheet and then compiled into a graph using the functions available through the software.

Age in months of students in our class

Age in months	Tally	Frequency
85–96	III	3
97–108	HHT HHT HHT III	18
109–120	HHT HHT HHT HHT III	23
121–132	HHT HHT HHT I	16
133–144	HHT II	7

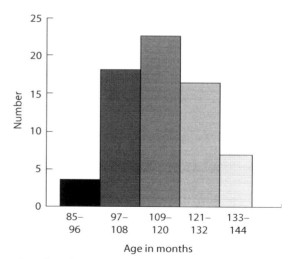

Age of students in our class

● Circle or pie graphs

Perhaps the most difficult graph for students to construct is the pie or circle graph. Considerable knowledge and skills need to be brought together in order to construct this graph—fraction knowledge, percentages, angles and operations. When the graphs are introduced, simple data should be used so that the fractions to be interpreted (and later drawn) are halves, quarters and thirds. As the students become more familiar with data representation, the teacher should ensure that the data can easily be transferred into a proportion of 360. For example, a sample of 18 will convert to an angle of 20°, which is an easy size for students to draw.

Informal introduction to circle graphs

Column graph to strip graph to create a circle graph

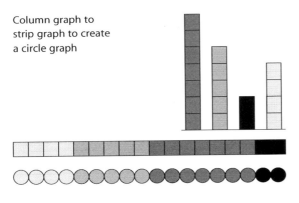

By the time circle graphs are introduced in the primary school, students will have had substantial experiences with column graphs. To introduce the idea of circle graphs, two strategies can be used. One of the easiest ways to begin constructing the graph is to have the data constructed as a single long strip. Once the graph has been constructed on paper, the columns are cut out and glued end on end to form a long strip. Once the long strip has been constructed, it is then joined to form a circle. The centre of the circle is located and the end of each section is then drawn to the centre.

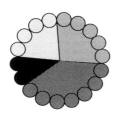

Using beads to construct pie graphs

A second method involves representing the columns through beads threaded on to a string. The items in a column are threaded on to the string. The string is then made into a circle. The centre is identified and lines are drawn from the end of each coloured section into the middle. These simple methods can be used for introducing the idea of circle graphs. They are, however, inaccurate, as considerable guesswork is involved in identifying the centre of the circle and whether the shape made is circular. The process introduces the construction of a circle graph, but without the complex construction process. This process allows the students to appreciate the purpose of the graph and to develop an understanding of the graph itself without being confused by the mathematics involved in the process. The purpose at this stage is to introduce the new graphical format and an appreciation of its role in representing data as a whole set, and the individual data sets as parts of that whole. This is a new idea for students so time needs to be spent developing it.

Once the students can construct circle graphs in this way, and have the requisite background knowledge in other areas of mathematics, they can construct the graphs using the traditional methods. Initially, the examples should be ones where the students can divide the circle into recognised sections such as halves, quarters, eighths, thirds, sixths or twelfths. The lines can be drawn from side to side (when using halves) or from the centre.

Other knowledges can be used to construct pie graphs. In particular, clock knowledge can be most useful. In the example here, the teacher has selected a sample with 60 items. Students can then apply their knowledge of minutes, quarters, halves (15, 30 and 45) to create appropriate divisions and then to divide the circle graph into segments to appropriately represent the data.

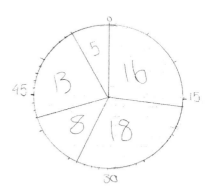

Student work sample: using knowledge of clocks to construct a pie graph

Formal construction of circle graphs

Constructing pie graphs where angles have to be calculated is generally left for the upper years of primary school, as understandings of so many other aspects of mathematics are needed. The examples developed by the teacher should be those that will compute easily into 360°. Early experiences need to focus on examples where students can divide the total number of items into 360° to find out the size of individual segments, and then multiply this by the number of items in each data set.

As with other teaching strategies, the teacher's role is to move the students to sounder methods of mathematising situations. Once students are confident with this simple method, a more general strategy can be developed so that they will be able to construct any circle graph, regardless of the numbers in the data set. Constructing tables such as those shown helps the students to systematically work through the long process. Rather than coming to rely on a formulaic thinking, students should be encouraged to see whether there is any connection between the size of the fraction and the degrees, in order to see whether the task can be calculated more easily.

Colour	No. parts	Size of selection
Blue	7	$7 \times 20 = 140$
Orange	5	$5 \times 20 = 100$
Red	2	$2 \times 20 = 40$
Pink	4	$4 \times 20 = 80$

Colour	Size of segment = percentage $\times$ 360
Blue	$7/18 \times 100 \times 60 = 140°$
Orange	$5/18 \times 100 \times 60 = 140°$
Red	$2/18 \times 100 \times 60 = 140°$
Pink	$4/18 \times 100 \times 60 = 140°$

Students exploring graphs to represent collected data

● Using technology

As with other areas of mathematics, the use of technology can greatly enhance students' understanding of circle graphs. The construction of circle graphs can be quite tedious, so using software packages such as spreadsheets enables students to enter data and construct the graphs they need much more rapidly.

● Stem and leaf plots

A stem and leaf plot is another way to represent data, and such a graph allows a greater appreciation of the scores or data set. A bar graph gives only the amount of scores or events within any particular column; there is no other information apparent. Consider a scenario where test scores are being plotted and have been grouped (as per a grouped frequency histogram). This type of graph would not tell us anything about the scores—for example, whether they were equally distributed across each score, or were clustered around particular points. You might get results back from an external system testing procedure and find that many of your students have scored in the 40–49 per cent range. Does this mean that they scored just below the pass mark, or were their scores much lower than this? A stem and leaf plot allows this information to be made apparent.

Much like the process that is undertaken for other graphs, the data must be collated and clustered. However, unlike the grouped frequency process, where only tallies are taken for any score, it is now necessary to keep track of the individual scores as they are clustered.

Consider the final scores for your class:

108	86	45	67	89	101
39	65	87	90	76	120
91	67	74	97	69	109
65	32	89	82	93	74

When these data are plotted on a column graph, it shows the spread of scores but its usefulness is restricted as the values of the actual scores are not evident. For example, there is no indication of whether the scores in any of the categories were high, mid-range or low.

A stem and leaf plot preserves the scores. Finding the range of scores, then plotting the stem in an upward growth is the general format.

120	0				
110					
100	8	1	9		
90	0	1	7	3	
80	6	9	7	9	2
70	6	4	4		
60	7	5	7	9	5
50					
40	5				
30	9	2			

Unsorted data showing scores

120	✓				
110					
100	✓	✓	✓		
90	✓	✓	✓	✓	
80	✓	✓	✓	✓	✓
70	✓	✓	✓		
60	✓	✓	✓	✓	✓
50					
40	✓				
30	✓	✓			

Tally marks do not show the actual scores

May								Stem	February				
							1	**12**	0				
							0	**11**					
						9	3	**10**	1	8	9		
9	4	3		1	0			**9**	0	1	3	7	
				9	7	3		**8**	2	6	7	9	9
		8	5	4	0			**7**	4	4	6		
	8	7	7	6				**6**	5	5	7	7	9
					1			**5**					
					0			**4**	5				
					4			**3**	2	9			

Sorted data using 2-sided representation

However, some graphs use only the tens place value in the stem and the ones value in the leaves. First, the data must be grouped as in the table shown. Once this has been done, the data must then be organised in order so that the smaller numbers are closer to the stem and radiate outwards. What can be observed quite easily with this type of graph is that there are numerous high scores and that within each scoring category students scored well.

The graph is also useful if two data sets need to be compared. For example, you may want to compare boys' and girls' results for the maths tests and compare the different outcomes; or compare the results from the two year levels that are in your class. In contexts beyond the classroom, it may be used for comparing scores for two teams over a season.

TEACHING IDEA

Book club

There are 7 students in the Ned Kelly reading group. If the mean number of books read by the group was 5 and the mode was 4, how many books could the students have read individually?

Note: This is a complex question and would be suitable for challenge work.

Using the same data set, the tests for a second test were produced and it seemed like the students would have scored much better had a simple column graph been used. However, the stem and leaf provides the detail to allow a comparison.

Measures of central tendency

One of the more common statistical measures encountered by students in their everyday lives is that of central tendency. Rather than focus on the calculation of various averages, current teaching also requires students to know when particular measures of central tendency are used and why; why they would report using such measures; the degree of accuracy needed; how different measures can be used to convince the reader, and so on.

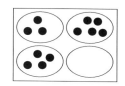

TEACHING IDEA

Mean, median and mode

In what situations would it be best to use:

Mean ....

Median ....

Mode ....?

● The mean

Mean refers to the arithmetic average where all scores are totalled and then divided by the number of addends. When teaching the mean, activities to support the development of the concept might include finding the mean height of the class. One way to do this is to measure lengths of streamer the same height as the students. These are taped end to end to form a single strip. This is then folded into the number of students in the group so that the mean height becomes apparent. Alternatively, using beads placed in an egg carton, students can represent the number of pieces of fruit they ate in one week. The corresponding number of beads is then placed in a hole representing each student, as shown in this example. The students then move the beads around until there are an equal number of beads in each hole, thus representing the mean.

Such experiences help students to understand the fundamental aspects of mean. Studies have shown that while students may be able to calculate means, they often have little understanding of the concept, even by the later years of middle schooling (Years 8 or 9), yet the concept is critical for everyday knowledge and problem-solving.

● The median

The median refers to the middle value in a set of data. In the data set of 2, 4, 5, 9 and 9, the median is the value that falls midway in the set, so the point is the third value—5. Teachers often use the analogy of a highway where the mid-point of the highway is the median strip, thus illustrating the notion of 'middle point'. This measure of central tendency is less well known to primary school students, but it is one with which they must become familiar since it is used often in reporting. Teaching this concept often takes place through the physical manipulation of data sets, where each score is written on an individual card and then organised in ascending or descending order. The end scores at each end are removed or marked off simultaneously so that there is only one (or two) cards left. The remaining card is the median. Where there are two scores, the mid-point between these scores is the median.

T E A C H I N G   I D E A

Consider the following range of salaries for staff within a new IT company:

$	$
35 000	42 000
31 500	76 000
28 000	120 000
37 000	41 000
39 400	46 500

What is the mean?
What is the mode?
What is the median?
Which is the best measure to report to the workers? Why?

● The mode

The mode refers to the score that occurs most frequently within a data set. When considering this value in relation to a column graph, it is the value with the highest column. It is usually easy to find and it is the one that is least affected by extreme values in a data set. The age of the students in a classroom is an excellent example of the application of mode, since in any one class there is likely to be a considerable number of students of the same age. A commonly reported mode in recent times is that of the ageing teacher population, where there are considerable numbers of teachers in their mid- to late forties. Calculating the mean age would not provide state authorities with a true picture of the depth of the issue, but when the mode is used it becomes clear that a large number of teachers are in this upper age bracket and that they are likely to retire within a few years, thereby creating particular problems for staffing.

Using the different measures of central tendency provides a useful forum for students to explore why different measures are used in different contexts, and why writers might prefer to use one measure over another, depending on the task at hand. This is an important life skill since the wider public needs to develop critical reading skills in terms of descriptive statistics in order to be able to analyse the ways in which texts have been used (rightly or wrongly) to persuade the reader. Much of the literature on critical numeracy focuses on the use and interpretation of statistics. Knowing how statistics can be used to empower and disempower readers is an important skill to be learnt.

Statistical literacy: Interpretation of data representations

While considerable time is spent teaching students how to construct graphs, less time is spent on teaching them how to interpret graphs, perhaps one of the most neglected aspects of the graphing sub-strand.

Where interpretation is taught, it tends to be very simplistic and does not prepare students adequately for the new forms of statistical literacy of the world beyond school. As noted earlier in this chapter, statistical literacy is a key life skill for contemporary times (Steen, 1999), and students need to leave school with this skill well developed. Much like our expectation that students will exit from school with appropriate levels of traditional literacy, so too they must now exit with the capacity to read the many types of data that saturate contemporary society. This means that they must have significant experiences in reading, interpreting, analysing and being able to critique the multiple forms of data they may experience. Data will be displayed through various media—including newspapers, the internet, television, brochures and radio. Understanding what is being said is critical in order be able to make informed decisions.

● Critical numeracy

Critical numeracy is a key skill related to statistical literacy. Interpreting data is often embedded within superficial learning environments. The deforestation of the Amazon, the emission of greenhouse gases from various countries, the amount of energy consumed by different countries, the water quality of the local creek, or the amount of government money spent on education in comparison with defence are all interesting data sets with which to work in the classroom. However, students need to work further with such data sets to explore questions of why such outcomes occur, and to look at what can be done to challenge some of the important issues that traditionally have been masked through statistics. For example, students may collect statistics about the water quality in the local creek and use spreadsheets to record and display their findings. Within a critical numeracy approach, questions are posed about the effects of the findings on the local ecosystems and, depending on the findings, students might decide to write a letter to their local member of parliament or take other social action to draw attention to the water quality. Critical numeracy seeks to empower students through allowing them to gain appropriate statistical knowledge and then enabling them to use that knowledge beyond the classroom setting.

Teaching notes for graphing

● Types of data

The two main types of data that can be collected—discrete and continuous—impact significantly on what can be constructed. While this aspect of data is not taught to students in primary school, teachers need to be very cognisant of it in order that they do not create misconceptions for students who may have to re-learn in their later years of schooling.

Discrete data

This type of data is that which falls into one category or another. This includes the gender of children (boy or girl); types of cars (Toyota or BMW); flavours of ice-cream (strawberry or chocolate); the grade of enrolment (Year 4 or Year 5), and so on. Data that belong to one category or the other are presented in *bar graphs*—graphs where the columns do not touch.

In the early years of schooling, data are often collected around themes that appear to be discrete, such as the season or the colour of hair or eyes. These categories may *appear* to be discrete since the season is either autumn or winter, or the hair is either black or brown, but this is not entirely accurate. A category such as a season may commence on the first of the month in some countries but on the equinox in other countries. Thus there is an arbitrary decision being implemented with regard to category.

Continuous data

The second category is continuous data. This type of data relates to information that exists along a continuum and at some point arbitrary categories are made. The data is usually something that can be measured—time, length, temperature, etc. When constructing graphs of ages of students in classes, categories of 5, 6 and 7 years might be used. This may appear to be discrete—you are either 5 or 6 or 7. However, on closer examination we can see that the ways in which the construct of 5 is made can vary. Is it when someone turns 5 and just before they turn 6? Is it somewhere between 4½ and 5½? Is it how old they are on a particular day? From 4 years 10 months to just before 5 years 10 months? Each of these categories means that the data concerning the students in a class can be organised in different ways. The reason here is that the construct of time is continuous: to make sense of 5 years for a category, we need to make decisions (and cut-offs along that continuum) as to how 5 will be defined. Since the data are continuous and the categories are 'slippery', depending on how the construct is defined, the corresponding column graph is a histograph (or histogram) where the bars or columns touch.

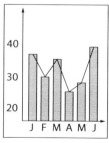

Rainfall in semester

The unique feature of histographs is that the bars are related to one another, and hence can be compared over a period. This means that this column graph can be converted to a line graph should the data need to be considered over time. This conversion is not possible with a bar graph.

● Purposes of graphs

In and of themselves, graphs are not entirely free choice in nature. The type of graph possible is influenced by the data that have been collected. When using graphs to represent data, the purpose of the graph is an important consideration, and different graphs serve different purposes. Graphs are used to display information quickly and easily so the right type of graph for the purposes of the display is important. For example, the class collected data on lunches since the unit of work being undertaken was Healthy Lifestyles with cross-curricular links. If the purpose of the task was to compare the different types of food items eaten, a column graph would be the most appropriate, since it would allow for quick reckoning on which lunches were eaten most frequently, least frequently, and so on. If the task was to consider what the class ate overall, a column graph would not allow this interpretation easily. A pie or circle graph would be far more useful in conveying this information.

● Technology, computers and spreadsheets

Just as the calculator has a unique role to play in developing students' number and computational sense, the use of computers (and particularly spreadsheets) has the potential to radically impact on the teaching of statistics. The construction of graphs has often been an arduous task, akin to the task of calculation in the number strand. The use of technology can support and enhance statistical and graphical understandings. Once students have developed a sense of the types of graphs, and their construction, interpretation and purpose, the task of constructing them can be replaced through the use of technological tools such as the spreadsheet. This frees up class time for more complex and meaningful activities.

Entering data into the spreadsheet, then making key decisions regarding the type of graph most useful for displaying the data, requires high levels of mathematical reasoning. This in turn allows a greater amount of teaching time to be spent on developing the appropriate

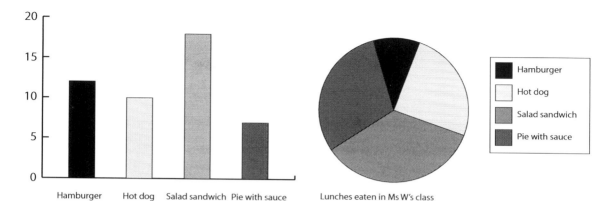

Lunches eaten in Ms W's class

Using spreadsheets
to construct graphs

thinking skills associated with construction and interpretation of data. Being able to estimate and evaluate whether the graph constructed is correct demands that students are familiar with the processes associated with data representation. Decision-making processes are informed by students' knowledge, and hence demand the appropriate levels of understanding.

Much like the debates associated with the use of calculators in number work, some sectors of the community may argue that the use of spreadsheets and other technology tools (such as graphic calculators) de-skill students. However, there is now a substantive body of research that counters this position (Asp et al., 1993; King, 1993; Vincent, 2000).

Teaching chance and data in the middle years

In this chapter, the importance of extensive experiences for promoting probabilistic reasoning and statistical literacy have been emphasised. The early years learning experiences must extend students beyond the 'birthday graph', to thinking about reasons why data are collected and what the data tell us, as well as engaging students in discussing chance outcomes. In the middle years, learning experiences must be opportunities for capitalising on promoting meaning for key probabilistic terms, such as randomness, variation and independence, and extending their critical numeracy skills associated with questioning data.

When students come to the middle years, they often carry with them misconceptions or naïve conceptions about chance events. They need to be immersed in a range of chance activities so that they can make conjectures and test them out. Common errors associated with the language of chance are highlighted when students are asked to describe the likelihood of an outcome when there are more than two possible outcomes and they respond 50/50 (Tarr, 2002; Watson, 2005). Quantifying events as 50/50 indicates students' awareness of measuring chance but not moving to the mathematical analysis required.

Another common misconception relates to randomness of sequences of outcomes and independence of events. When a coin is tossed three times and it has come up tails all three times, many students will state that the next one will be heads because 'it is the heads' turn'. Because calculation of mathematical probabilities (theoretical probability) does not match actual outcomes (experimental probability), students need to reconcile experimental and theoretical probability outcomes by building understanding of the notion of 'in the long run'. Real-world examples can be drawn upon to stimulate discussion about theoretical and experimental probability outcomes in relation to insurance premium calculations, football betting, horse racing, health outcomes, and so on. Making probability real and relevant promotes students' appreciation of probability and its direct application to our daily lives.

A further common misconception is in relation to populations and sampling. Research shows many middle years students are unsure about representative sampling. When asked about the best method to survey students in a school by either (a) putting all 600 students' names in a hat and drawing out 60, (b) surveying 60 friends, or (c) asking volunteers to participate in the survey as they approach a specific location in the school year (canteen), only 12 per cent of Year 7 to 12 students recognised choice (c) as a potential source of bias, and one-third of students thought (b) was a good method (Shaughnessy, 2007). To promote statistical thinking in middle years, population and sampling is important. Other key teaching points in relation to statistics, as suggested by Shaughnessy (2007) are as follows:

- *Emphasise variability*—rather than just describing sets of data in relation to mean (median and mode) scores, the outliers are also important to consider the variability in the data.
- *Compare data sets*—students should be immersed in a range of opportunities to collect data that they analyse 'by eye' (rather than

determining measures of central tendency—mean, median, mode and other data analysis methods).

In relation to probabilistic reasoning, the big ideas are: 'chance variation, randomness, independence (and their complementary elements stability, regularity and co-occurrence)' (Jones et al., 2007).

■ REVIEW QUESTIONS

14.1 List the general sequence for teaching graphs. Provide illustrations of the sequence.

14.2 Discuss the difference between theoretical and experimental probability.

14.3 When is it more useful to use theoretical probability as opposed to experimental probability in the teaching of probability? Provide illustrative examples to justify your argument.

14.4 Students often develop misconceptions about the independence of events. Select a situation where the events are independent and discuss how you would go about teaching this concept so as to overcome misconceptions.

14.5 Identify particular instances of when different measures of central tendency are more appropriate than others.

14.6 Discuss the value of using technology to support learning.

Further reading

Aspinwall, L. and Shaw, K. (2000). Enriching students' mathematical intuitions with probability games and tree diagrams. *Mathematics Teaching in the Middle School,* 6(4), 214–20.

Callingham, R. (2000). Come in spinner! Innovative assessment in the primary Classroom. *Australian Primary Mathematics Classroom,* 5(1), 9–14.

Feicht, L. (1999). Making charts: Do your students really understand the data? *Mathematics Teaching in the Middle School,* 5(1), 16–18.

Mulligan, J. and Bobis, J. (1998). Investigating data exploration. *Reflections,* 23(1), 19–22.

Perry, B. (1998). Frameworks for statistical thinking in primary schools. *Reflections,* 23(1), 85–8.

Quinn, R. (2001). Using attribute blocks to develop a conceptual understanding of probability. *Mathematics Teaching in the Middle School,* 6(5), 290–4.

Ritson, R. (2000). A question of choice. *Australian Primary Mathematics Classroom*, 5(3), 10–14.

Stephens, M. (1998). Using data representations to ask hard questions. *Australian Primary Mathematics Classroom*, 3(4), 23–7.

Tarr, J. (2002). Providing opportunities to learn probability concepts. *Teaching Children Mathematics,* 8(8), 482–7.

Wiest, L.R. (1998). In the face of uncertainty. *Australian Primary Mathematics Classroom*, 3(4), 19–22.

References

Asp, G., Dowsey, J. and Stacey, K. (1993). Teaching mathematics with technology: Computer spreadsheets and graphing applications. In J. Mousley and M. Rice (eds), *Mathematics: Of primary importance* (pp. 86–92). Melbourne: Mathematics Association of Victoria.

Dixon, J.K. and Falba, C.J. (1997). Graphing in the Information Age: Using data from the World Wide Web. *Mathematics Teaching in the Middle School*, 2(5), 298–304.

Jones, G., Langrall, & Mooney, E, (2007). Research in probability: Responding to classroom realities. In F. Lester (ed.), *Second handbook of research on mathematics teaching and learning* (pp. 909–56). Charlotte, NC: Information Age.

King, M. (1993). Data handling using spreadsheets in primary schools. In J. Mousley and M. Rice (eds), *Mathematics: Of primary importance* (pp. 80–5). Melbourne: Mathematics Association of Victoria.

Luke, A. and Freebody, P. (2001). *Report of the literacy review for Queensland state schools.* Brisbane: Education Queensland.

Shaughnessy, J.M. (2007). Research on statistics learning and reasoning. In F. Lester (ed.), *Second handbook of research on mathematics teaching and learning* (pp. 957–1010). Charlotte, NC: Information Age Publishing.

Steen, L.A. (1999). Numeracy: The new literacy for a data-drenched society. *Educational Leadership*, October, 8–13.

Tarr, J. (2002). The confounding effects of '50–50 chance' in making conditional probability judgments. *Focus on Learning Problems in Mathematics*, 24, 35–53.

Urso, J. (1999). What do you like to read? *Teaching Children Mathematics*, 6(1), 34–7.

Vincent, J. (2000). *Computer enriched mathematics for Years 5 and 6.* Melbourne: Mathematics Association of Victoria.

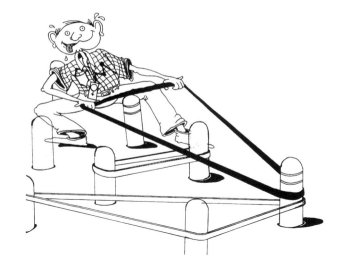

CHAPTER 15

Space

The topic of space includes developing students' knowledge about aspects of geometry such as two- and three-dimensional shapes, and notions of symmetry, transformation and tessellation. Space also includes learning about position, location and arrangement in relation to maps, plans, scaled drawings and grid systems. Spatial knowledge extends from learning the simple language of position and arrangement in the early years—inside, outside, near, far, etc.—to using a map to determine distances in the later years. Also important in the topic of space are developing students' spatial visualisation and spatial reasoning abilities, as well as making links to natural and built environments. Learning activities in this topic can involve exploration and investigation of buildings, bridges, consumer products, shapes and symmetry in plants and animals, and shape and function of natural and manufactured (constructed) objects. Everyday tasks such as map-reading and furniture arrangement, and the construction of items from plans, involve spatial reasoning. The space topic also links to many occupations—architecture, landscape gardening, golf course designing, civil engineering, surveying, town planning, painting, crafts, art, building and dress designing. It is important to make connections from space to measurement (Battista and Clements, 1998; Trafton and Hartman, 1997) and number topics, and to other curriculum areas such as art, craft, physical education, science and social studies (Morrow, 1991).

Topics in the space strand

Topics within the space strand include location (position, direction, coordinates), networks (maps, scale drawing, topology), three-dimensional solids (surface, corner, edge), two-dimensional shapes (sides, angles, polygons), symmetry (line symmetry, rotational symmetry), tessellations (tiling, design, art applications), dissections (puzzles), similarity (shape, perspective) and transformations (flips, slides, turns). Many of these topics are interrelated, and thus the space strand can be organised in different ways. Here we have organised the topics under shape and structure; transformation and symmetry; and location and arrangement.

Visualisation

Being able to visualise images is an important foundation for problem-solving in mathematics (Owens et al., 2001). Developing a strong facility for visualisation will result in an increased capacity to solve problems. The development of visualisation is fostered through the study of space, so developing spatial sense is very important (Del Grande, 1990).

The study of geometry and space assists the development of visual processing—the capacity to create, manipulate and transform spatial images in the mind. The exercise in the diagram requires students to be able to visualise the rectangular image and rotate it in order to respond to the question. Visualisation also assists in interpretation of information represented in diagrams, graphs and maps. Lowrie and Diezmann (2007 and 2009) have found that graphical literacy—the capacity to read diagrams, maps and graphs—is an important component of the space strand but it is a skill that needs to be developed, particularly from around the middle years of schooling—that is, from Grade 4 onwards. Lowrie and Diezmann also report that in the middle years, there are gender differences in graphical literacy of which teachers also need to be cognisant, and it is important to develop strategies to enhance these skills and redress differences.

Visual processing is a key skill in mathematical problem-solving across the strands. It is recognised that the teaching of spatial concepts and processes is the key to students developing visualisation and spatial sense (Pegg and Davey, 1989). The most effective problem-solvers

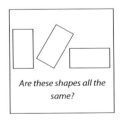

Are these shapes all the same?

Creating, manipulating, transforming images

have a strong visual sense as well as verbal, symbolic and abstract thinking. Throughout their study of space, students should continually be challenged so that they do not form inflexible images. An example of an inflexible image is when students correctly identify a square when it is aligned horizontally but insist it is a diamond when it is tilted on one corner. Students holding inflexible images often cannot identify forms other than an equilateral triangle as triangles. In a similar vein, triangles with their bases uppermost may be regarded as 'wrong' or 'upside down'. Thus, in order to enhance visualisation skills, early shape-recognition activities must provide students with a range of forms and a variety of orientations for each particular shape.

The van Hiele levels of geometric thinking

The most cited and used theory underpinning the teaching of geometry is that proposed by Dutch researchers van Hiele and van Hiele (Pegg, 1985), who note that students' geometric thinking typically can be represented by key characteristic styles, which they refer to as 'levels'. According to this theory, students' progression across the levels occurs in response to instruction, emphasising the impact of teaching on students' thinking. (The levels are not intended in the sense of broad developmental or biological progression.) The theory provides a useful model for planning teaching (Southwell, 1998). The model has five levels, but for the primary school teacher the first three are most relevant.

Level 1: Visualisation

At Level 1, students' thinking about shapes relates solely to what they see. Students react to figures as wholes—a square is a square because it is a square. Similarly, an angle is an angle because it is an angle. Students do not think about the components of a shape or its properties. They might regard a square that has been rotated through 45° as a diamond—that is, they no longer see it as a square.

Level 2: Analysis

At level 2, students begin to focus on the properties of shapes and to informally analyse the components of shapes—a square is a square because it has four equal sides and four right angles. However, they do not focus on relationships between properties.

Level 3: Abstraction or informal deduction

At level 3, students logically order properties, form abstract definitions and identify necessary and sufficient conditions associated with properties. For example, a square has some properties related to other shapes—a square can be both a rectangle and a rhombus; if the opposite angles of a quadrilateral are equal, then the opposite sides are parallel; if I know that a parallelogram has one right angle, then it must be a rectangle; a triangle with three equal sides will have three equal angles.

Level 4: Deduction

Level 4 involves formal reasoning, where axioms, terms, underlying logical systems, definitions and theorems become the focus. This level is commonly associated with high school geometry, and includes developing formal arguments and proving theorems.

Level 5: Rigour

Level 5 involves comparing systems based on different axioms—for example, Euclidean geometry versus spherical geometry. Typically, the study of geometry associated with this level of thinking is undertaken beyond the high school years.

● The importance of the van Hiele model

Central to the van Hiele model is that progression in geometric thinking is due to effective teaching, and is not necessarily associated with increasing age or maturation—effective teaching is the essential element. By focusing on the different ways students think geometrically, the van Hiele model allows teachers to identify how students are conceptualising geometric ideas and then how best to plan and organise effective learning so that the student can progress to more advanced thinking. For example, a student working at level 2 may not be able to see that a square is a special kind of a rectangle. By recognising that the student is not seeing interconnections between properties, the teacher is able to organise appropriate learning experiences that have the goal of enabling the student to see connections between the properties of rectangles and of squares. In this way, students can become aware that a square is a special kind of a rectangle—a rectangle with the additional property that its adjoining sides (as well as its opposite sides) are equal.

For students in the primary years, the first three levels of the van Hiele model are generally very applicable. Using the model to carefully plan teaching can lead to enhanced understandings of geometry by students.

Teaching notes

The research of van Hiele and van Hiele has informed much of the current teaching of space. Central to their work is the important role of teaching of geometric ideas. They argue that the complexity in students' geometric thinking arises from the experiences provided by teachers. Space activities should link to and draw from the everyday world and build on students' experiences with spatial aspects of their environments. Space activities should be multisensory, informal before formal, and provide ample opportunities for the development of visualisation.

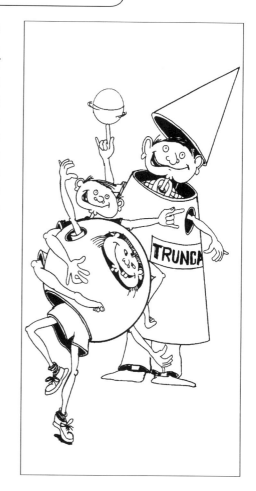

Appropriate geometric activities should use and develop important skills such as:

- visualising—manipulating and transforming spatial images in the mind; interpreting figural information; interpreting and reading diagrams or maps
- communicating—developing accurate and precise geometric language, both spoken and written (start with children's natural language and gradually develop more formal terminology)
- drawing and modelling—sketching and modelling with appropriate accuracy

- thinking and reasoning—asking students to give reasons for their answers: Why did you say that? How do you know that? What evidence is there for that?
- applying geometric concepts and knowledge—drawing attention to real-life applications; deriving problems from real-life situations; considering relationships between shape and function.

Spatial concepts: Early activities

Play activities in the early years are important for developing spatial concepts. Rather than being random, play activities should be planned carefully and their purpose should be clear. Early activities include:

- sorting—according to students' criteria or criteria suggested by the teacher; for example, rolling or sliding, height, pointiness; games of identifying which one is different
- building—Lego, blocks, junk; making the tallest or widest structure; making a symmetrical building; making an enclosure for this object
- packing—placing materials neatly into storage boxes
- modelling—with plasticine, clay, straws and string, cardboard and tape or rubber bands, toothpicks
- touching—using touch to identify a shape (feely bag); describing the object to others
- matching—shapes to diagrams; matching a picture and a model; building a three-dimensional structure from its picture; matching containers with models
- drawing—solids from different viewpoints; identifying and drawing shapes after walking through the park, the playground, the garden; discussing reasons particular shapes have been used for various structures
- visualising—replicating structures behind a screen that have been shown for a few seconds.

Shape and structure

● Two-dimensional shapes

The study of two-dimensional space relates to topics such as polygons, lines and angles. Polygons are two-dimensional shapes with three or more sides. The word polygon comes from two words: *poly*, meaning many, and *gon*, meaning sides. Hence a polygon is a many-sided figure. Polygons are also regarded as plane shapes (or figures), as they lie on the two-dimensional plane. The name of a specific polygon includes a prefix determined by its number of sides (see table).

Names of polygons

No. of sides	Prefix	Name
3	tri-	triangle
4	quad-	quadrilateral
5	penta-	pentagon
6	hexa-	hexagon
7	hepta-	heptagon
8	octa-	octagon
9	nona-	nonagon
10	deca-	decagon
. . .	. . .	. . .
many	poly	polygon

Polygons are classified as regular or irregular. Regular polygons have all sides equal and all angles equal. An irregular polygon has either some or all of its sides of different lengths, or some or all of its angles of different sizes, or both sides and angles differing. In some cases, students might be presented with polygons that are almost exclusively regular. Limiting students' experiences in this way can result in their having an impoverished notion of polygons. They might have difficulty in naming certain polygons. Thus it is important that students engage in activities with both regular and irregular polygons.

In the early years, the focus is on assisting students to identify common shapes such as squares, triangles and circles (although a circle is not a polygon. Any shape with curved sides is not a polygon). Richer conceptual understanding is promoted through activities that direct students to consider properties of shapes, and by providing a variety of examples of particular polygons—both regular and irregular—in various spatial orientations (Fox, 2000). Sorting activities are useful for preventing the development of inflexible conceptualisations of particular shapes.

When teaching basic geometric shapes, students need to become familiar with shapes presented in a range of orientations. If students' experience of two-dimensional shapes is mainly limited to shapes in an upright orientation (i.e. where the base line is on the horizontal orientation), this can result in misconceptions. For example, students may regard only one of the two squares in the diagram as a square. Therefore it is important that students are presented with examples of basic 2D shapes in a variety of orientations.

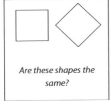

Are these shapes the same?

Visualising shapes

Tangrams are inexpensive and useful in the teaching of two-dimensional shape. Tangram activities not only support the learning of the names of shapes but also help students to develop spatial visualisation. Various tangram puzzles are available but a seven-piece tangram is most commonly used (van Hiele, 1999). The tangram allows students to make many shapes, both prescribed and free form. It also allows them to explore aspects of transformational geometry as they create or copy shapes.

7-piece tangram

T E A C H I N G   I D E A

Geoboard activities

Exploring lines

- Make the shortest line segment possible on your geoboard.
- Make the longest line segment possible.
- Make three line segments of different lengths.

Exploring angles

- Make a narrow angle.
- Make a wide angle.
- Make an angle like the corner of a square.
- Make an angle with 2 nails between the rays.
- Make an angle with 0 nails between the rays.

Exploring regions

- Construct a region that has just one nail inside it.
- Construct a region that has 0 nails inside it.
- Construct a region that has 1 nail outside it.
- Construct a region that has 1 nail outside and 3 nails inside it.
- Construct a region that has 3 nails on the boundary.

Line and angle

Learning about line and angle requires students to be engaged in constructing a variety of lines and angles and discussing them.

Geoboards are a valuable resource for investigating lines and angle. Geoboards are wooden (or plastic) boards with nails protruding in a grid pattern. Frequently they are in a square array, but can be in circular patterns as well. A variety of two-dimensional shapes can readily be created on a geoboard using rubber bands, and language associated with shapes, lines and angles can be fostered (Britton and Stump, 2001).

The geoboard is also a valuable resource for constructing and learning about angles. Many students develop misconceptions about angles (Mitchelmore, 2000), particularly in relation to the size of the arms of an angle versus the size of an angle. The notion of angle relates to measuring an 'amount of turn'. A simple way to introduce students to the idea of angle is to ask them to stand with their arms outstretched in front of them and then open one arm to the side through an angle of 90°. This can be linked to turning all the way around, which is a turn through 360°, and turning halfway around, which is a turn of 180°. Outstretched arms can be dropped to the side of the body and turning can continue in order to emphasise that the notion of angle relates to an *amount of turn* rather than the length of the arms or the length of the lines making an angle.

TEACHING IDEA

Sorting shapes

Which of these shapes are triangles? Put a cross on the shapes that are triangles.

Cut out the shapes and sort the triangles into one group. Explain why you sorted the shapes out in that way.

To focus further on angle (e.g. the angles made by the hands of an analogue clock) an angle tester can be created. This is simply a circle of paper folded into four quarters. The corner created is a right angle and can be used to find angles that are exactly 90°, less than 90° and greater than 90°. An angle wheel (see diagram) can be constructed using two circles of equal size cut from different coloured pieces of paper. By cutting a split to the mid-point of both circles, fitting the circles together and turning them, parts of each circle can be seen, from either side, forming angles of different sizes. Angles can be classified as obtuse (blunt), acute (sharp), right angled (like the corner of a square) and straight (flat line).

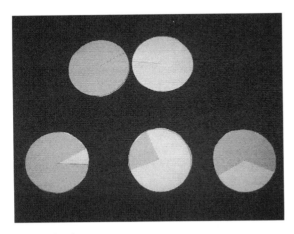

Angle wheels

● Three-dimensional shapes

The topic of three-dimensional (3D) space focuses on learning names and properties of solid figures (3D objects), relationships among solid figures and visualisation skills associated with this topic. Learning about solid figures can include learning about 3D shapes that are important in geometry (e.g. cubes, spheres, pyramids, prisms, cylinders and cones), and linking this knowledge to shapes and objects that occur in the natural and built environments. For example, a block of wood might have the shape of a prism, a basketball has the shape of a sphere and a funnel has the shape of a cone. Activities involving the construction of solid figures can support students' development of spatial knowledge (Ambrose and Falkner, 2002).

Many syllabuses list the topic of 3D space before the topic of 2D space. This is because young students can think more easily about solid figures than they can about plane (two-dimensional) figures. Solid figures can be regarded as more experientially real for young students because they are essentially objects that can easily be handled whereas the notion of a plane figure (2D shape) is more abstract.

Naming and classifying solid figures

It is important to understand the nomenclature of solid figures. There are parallels between the nomenclature of solid figures and that of plane figures, and teachers should be aware of the links in these two naming systems. The term 'solid figure' is used in an inclusive sense—any 3D shape is a solid figure. Shapes in the sub-group of solid figures that have flat surfaces (faces) only, are referred to as 'polyhedra' (singular 'polyhedron') as the word polyhedra comes from two Greek words: *poly*, meaning many, and *hedra*, meaning faces. Just as circles and semi-circles are plane figures that are not polygons, so spheres, cylinders and cones are examples of solid figures that are not polyhedra (because of their curved surfaces). Examples of polyhedra are cubes, rectangular prisms and square-based pyramids.

Polyhedra (prisms and pyramids) and solids (cones and cylinders)

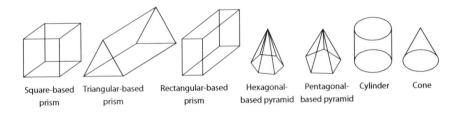

Square-based prism   Triangular-based prism   Rectangular-based prism   Hexagonal-based pyramid   Pentagonal-based pyramid   Cylinder   Cone

Prisms and pyramids

Look at the diagram of solid shapes above. Compare the prisms with the pyramids. What do you notice? The first thing you will notice is that the pyramids form a point. Pyramids are named by the two-dimensional shape of their base. All other sides of the shape are triangles. Prisms have opposite faces of particular polygons (e.g., square, triangle, pentagon) and all other faces are rectangles.

TEACHING IDEA

Sorting solids

Which of these shapes roll and which of these shapes don't roll?

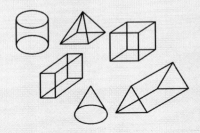

Nets of solids

The notion of a 'net' is the basis of an important activity in space and geometry that helps to connect ideas in 2D and 3D space. The net of a solid is a plane figure that can be folded up to make a hollow representation of the solid. Early learning activities for nets commence when students take a cardboard box (such as a biscuit box) and unfold it. They can see how the 3D biscuit box was constructed from a flat 2D shape. The specific shapes that were used to construct the 3D box can readily be seen. Students can then be challenged to create their own net to create a new box. More advanced activities can involve students drawing nets for a range of solids, then using them to construct the solids. Keep in mind that nets can be made for some solid figures that are not polyhedra (e.g. cylinders and cones). It is important for students to realise that a net is a flat, 2D shape that is folded to create the 3D shape. Even though the net can be seen to consist of several 2D shapes, they are connected together in one flat shape ready for folding.

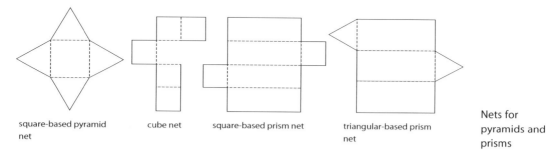

square-based pyramid net   cube net   square-based prism net   triangular-based prism net

Nets for pyramids and prisms

Solid, hollow and skeletal forms

Using a wide array of resources is invaluable in terms of creating opportunities for students to construct understandings of solids and their properties. Seeing the same solid shape in three different forms (solid, hollow and skeletal) is a key teaching strategy to develop deeper understandings of 3D shape properties. Teaching aids for 3D space typically include a set of solid figures made of wood. Hollow figures—that is, figures made from nets or from connecting plane figures—enable students to focus on the number of flat surfaces of the shape—that is, its faces. Skeletal figures can readily be made by using straws or toothpicks connected with marshmallows or Blu Tack. These shapes promote familiarity of language for 3D shapes of vertices, faces and edges. When making hollow shapes, students are working with each

Students making skeletons with straws and Blu Tack

Solid, hollow and skeletal forms of a cube

separate flat surface of the shape—that is, its faces. When making skeletal models, the 'skeleton bones' clearly highlight the edges of the shape and the connectors (marshmallows or Blu Tack) emphasise the vertices (pointy bits).

Euler's formula

To expand students' understanding of 3D solids, the exploration of Euler's formula provides an interesting investigation. Euler (pronounced Oiler) was a mathematician of the eighteenth century. Provide students with a range of 3D polyhedra (remember, 3D solids with curved surfaces are not polyhedra so don't include cones and cylinders) and have students create a data table of the number of vertices, faces and edges of each of the shapes. After they have completed their table, ask them whether they can see a relationship between the number of vertices, faces and edges. They will see that, for each shape, the number of vertices plus the number of faces is equal to the number of edges minus 2 ($V + F = E - 2$).

	Vertices	Faces	Edges
Cube	8	6	12
Rectangular prism	8	6	12
Square-based pyramid	5	5	8
Triangular-based pyramid	4	4	6
Triangular-based prism	6	5	9

Transformation and symmetry

Transformation and symmetry are interrelated topics. Transformation, as the name implies, is about altering a shape in some way. This could be altering its position in space (congruent transformation), changing its size (projective transformation) or changing its features (topological transformation). The terms 'object' and 'image' are often used in association with transformation. The figure prior to the transformation is referred to as the object; after the transformation it is referred to as the image. Symmetry includes line symmetry and rotational symmetry in 2D space, and plane symmetry and rotational symmetry in 3D space. The topic of tessellations (or tiling), is an aspect of 2D space that draws on notions of both transformation and symmetry.

● Congruent transformations

A shape transformation that leaves the original shape unchanged, but alters its position, is called a rigid or congruent transformation. A rigid transformation occurs when a shape is rotated, reflected or translated from its original position. These transformations are often referred to as flips, slides and turns. In congruent transformations, the image is always congruent with the object. Slides preserve the directions of lines, and both slides and turns preserve orientation. The transformations of flip, slide and turn relate closely to the topics of symmetry and tessellations (see below).

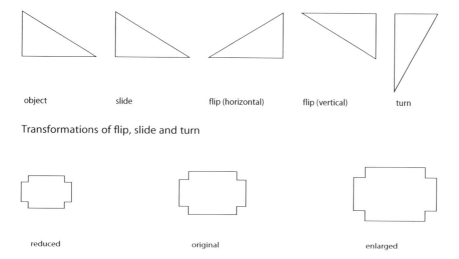

object   slide   flip (horizontal)   flip (vertical)   turn

Transformations of flip, slide and turn

reduced   original   enlarged

Transformations of enlargement and reduction

● Projective transformations

Projective transformations are enlargements or reductions of the original object. An enlargement is a transformation where the object is made larger while maintaining the original shape. A reduction is a transformation where the object is made smaller while maintaining the same shape. Thus enlargements and reductions preserve shape, but not size. They also preserve directions of lines and orientation. Projecting shapes on to a wall and tracing the image, and then comparing it with the original object, provide a great way to introduce studens to this type of shape transformation. Shadows can be linked to explorations of projective transformations.

● Topological transformations

Topological transformations involve stretching and bending without breaking or tearing. Properties such as lengths of lines, sizes of angles and straightness are ignored. An object and its image under a topological transformation are said to be topologically equivalent. Thus a quadrilateral, for example, is topologically equivalent to any other polygon. A soccer ball that is not inflated is a flat shape. When inflated, it takes on the (spherical) shape of a soccer ball. The flat shape and the air-filled shape are said to be topologically equivalent.

A lump of plasticine is an ideal starting point for exploring topological geometry. Rolling the plasticine into a sphere, then changing its shape into a cube demonstrates that these two shapes are topologically equivalent. Other shapes made by manipulating the plasticine—a bowl, a plate—are also equivalent. However, if a hole is put in the plasticine, the shape is no longer topologically equivalent. Thus, for example, a sphere and a doughnut shape are not topologically equivalent (see 'Teaching idea—Topological shapes').

TEACHING IDEA

Topological shapes

Shapes can be classified topologically by the number of holes in their surface.

- Start with a lump of play-dough with no holes in its surface. Make it into a ball shape. Now make it into a cube shape.
  1. Did the amount of play-dough change from what you started with?
  2. Are the shapes topologically the same?
- Start with a lump of play-dough and put a hole right through it (use a pencil). Make a doughnut and then the number 9 out of play-dough.
  1. Does each shape have just one hole?
  2. Did the amount of play-dough you started with change?
  3. Are the shapes topologically the same?
- With your play-dough, make the digits 0 to 9. Classify each one topologically.
- Find out if all the letters in your name belong to one particular topological category.
- Is it possible to have a given name in which all the letters are from one topological category?

The Möbius strip can be the basis of a rich topological investigation. Constructing a Möbius strip involves taking a long strip of paper, turning one end through one half-turn, and taping the two ends together. If a line is drawn down the middle of the Möbius strip, it eventually meets its beginning point. Thus a Möbius strip is considered to have one side only. Interesting activities can begin with making a longitudinal cut about one-third in from the edge of the strip (see 'Teaching idea—the Möbius strip'). Making cuts at other distances in from the edge of the strip yields different results. These activities can lead to conjectures, hypotheses, discussion and questioning.

The Möbius strip

1. Make a Möbius strip by cutting out a flat strip of paper about 50 cm long. Twist the paper (once), then join the two ends to make a closed ring.
2. Try colouring one side of the strip red and the other side green (or any two different colours). What do you notice?
3. Try to draw a line along the centre of the strip, continuing until you come back to the same point from which you started. Are you convinced that this strip has only one side?
4. Predict what you think will happen if you cut along the middle of the strip of paper. Try it. Were you surprised by the result?
5. Make another Möbius strip. This time draw a line along the strip one-third in from the edge. Continue to draw the line the same distance from the edge until you return to the point from which you started. Cut along the line you have just drawn. What happened? Did you expect that would occur?
6. Make a Möbius strip which is twisted twice before the ends are joined. Repeat activities 3, 4 and 5.
7. Make Möbius strips with three or four twists and repeat activities 3, 4 and 5. Can you discover a pattern emerging?

● Symmetry

Symmetry is a geometrical property that can be identified in everyday objects in both the natural and built environments as well as in some geometric figures (2D and 3D shapes).

Line or lateral symmetry

trapezium with one line of symmetry

arrow with one line of symmetry

rectangle with two lines of symmetry

pentagon with five lines of symmetry

Links to symmetry

Symmetry can arise in several ways. The most easily recognisable kind of symmetry is that associated with 2D space and the idea of a mirror image. This kind of symmetry can be seen in natural objects such as leaves and in geometry in plane figures (2D figures). As seen in the examples below, symmetry of this kind is associated with a *line of symmetry*. This kind of symmetry is referred to as line symmetry, lateral symmetry or mirror-image symmetry. An important activity for students is to use a mirror to reflect images of objects back on themselves in order to determine lines of symmetry. This activity can be applied to everyday flat objects (leaves), to pictures (a person's face), to symbols such as many upper-case letters, and to 2D figures. Place the mirror on the centre line of an object and see whether the reflected half of the image fits exactly on the actual object. Line symmetry can also be investigated through paper-folding along a line. To test whether a line of symmetry exits, cut out the shape out of paper, and fold it in half back on itself along the predicted line of symmetry. If it folds back exactly, that is the line of symmetry.

Try to categorise the 26 upper-case letters according to the number of lines of symmetry. Students have opportunities to determine all possible lines of symmetry for a range of 2D shapes—various triangles and quadrilaterals (square, rhombus, kite, etc.), and various pentagons, hexagons, etc. There is also a correspondence between line symmetry and the congruent transformation of a shape by flip or reflection (see above). When a shape is flipped, it becomes a mirror image of the original object. A line is drawn to indicate the exact point where the flip occurred to result in the mirror image. This is the line of symmetry for an object under a flip transformation. The figure on one side of the line of symmetry is the image of the object on the other side.

Rotational symmetry in 2D space

Symmetry in 2D space can arise in two ways—line symmetry and rotational symmetry. Rotational symmetry is associated with turning the object or 2D figure through a unit fraction (12, 13, 14, etc.) of a full circle. To test for rotational symmetry, draw an outline of the shape on another piece of paper. Place the shape on its outline. Mark the 'top point' of the shape with a dot. Turn the shape until it completely matches its outline. Count the number of times it fits its outline until it reaches the original position. An object or figure is said to have rotational symmetry of order two if, when turned through one-half of a circle (180°), the image corresponds exactly to the object in terms of the position it occupies. Rotational symmetry of order two is the simplest case of rotational symmetry and is referred to as point symmetry. In a similar vein, an object or figure is said to have rotational symmetry of order three if, when turned through one-third of a circle (120°), the image corresponds exactly to the object in terms of the position it occupies. Rotational symmetries of orders four, five, six, etc. are defined similarly.

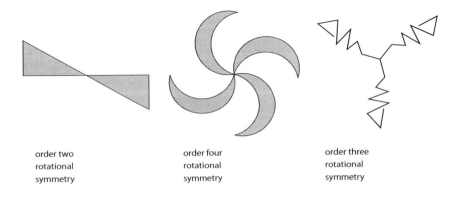

order two
rotational
symmetry

order four
rotational
symmetry

order three
rotational
symmetry

As in the case of line symmetry, activities relating to rotational symmetry can involve flat objects, symbols (upper-case letters) and 2D figures. Try to categorise the 26 upper-case letters in terms of whether or not they have rotational symmetry and, if so, of what order. A final point to keep in mind is the separateness of line symmetry and rotational symmetry. An equilateral triangle has three lines of symmetry and rotational symmetry of order three, and a square has four lines of symmetry and rotational symmetry of order four. However, line symmetry and rotational symmetry are not always associated in this way. Try to draw figures that have rotational symmetry of various orders and no lines of symmetry, or figures with one or more lines of symmetry but no rotational symmetry.

Symmetry in 3D space

Symmetry can also be considered in 3D shapes. Imagine cutting a shape in half. Do you have two identical shapes? If yes, then the cut that yielded the shape to mirror itself is called a plane of symmetry. Consider a solid figure such as a cylinder—how many planes of symmetry does it have? Since circles have an infinite number of lines of symmetry, cylinders have an infinite number of planes of symmetry. Rotational symmetry can also be considered in 3D space. Rotational symmetry about a point in 2D space corresponds with rotational symmetry about an axis in 3D space. As in the case of planes of symmetry, students could investigate axes of symmetry, and the order corresponding to each axis, associated with 3D figures and other everyday solid objects.

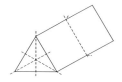

TEACHING IDEA

Symmetry in 3D space

- Find the planes of symmetry in the five regular solids.
- Identify some everyday objects—such as boats, cars, caravans, etc. that may have planes of symmetry.
- Classify objects as symmetrical and non-symmetrical.

● Tessellations

The topic of tessellations is associated with 2D shape and links to the everyday notion of tiling. A 2D shape is said to tessellate if it can cover an area without gaps or overlaps. Below are some examples of tessellating shapes. They extend in all directions without gaps or overlaps.

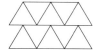 

Tessellating shapes

The following examples do not tessellate as there are gaps when the shapes are placed together.

Non-tessellating shapes

Tessellations

- Which of the regular polygons tessellate? Think about each of these in turn—equilateral triangle, square, regular pentagon, regular hexagon, etc.
- Consider a scalene triangle, e.g. with sides of 6 cm, 8 cm and 13 cm. Will this triangle tessellate?
- Consider a quadrilateral with sides 4 cm, 7 cm, 9 cm and 12 cm. Will this figure tessellate?
- Try to find a pentagonal shape that will tessellate.

One of the appealing aspects of the topic of tessellation is its potential for integrating 2D geometry and art. Use of tessellation in artistic and decorative contexts has a long history. This topic lends itself well to investigations of library sources and websites for artistic examples of tessellations as they occur in a range of cultural and historical contexts (Paznokas, 2003). Investigations could include finding information about, and examples of, the work of Martin Escher, who embraced both art and geometry. His work included unusual and picturesque examples of tessellation (see 'Teaching idea—Escher tessellations'). Students can use tessellation as a basis for their own artwork.

Escher tessellations

Take a regular tessellating shape such as a rectangle.

Make a cut from one end and then slide this to the opposite end and glue. Do the same to the corresponding edges.

This odd shape may not appear as if it will tessellate. Students should be allowed to predict whether it will tessellate or not. Provide opportunities for students to create their own tessellations and share with the class.

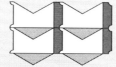

To create Escher-type artworks, begin with a shape that tessellates. Change the shape of one or more sides—for example, by introducing curves and using congruence transformations—flips, slides or turns—as the basis for additional and compensating changes that preserve the tiling aspect of the shape. In this way, the topic of tessellation is seen to link closely to the topics of transformations and symmetry (Whitaker, 2001).

● Technology and geometry

Several computer software programs have been developed to support teaching and learning in geometry. LOGO®, for example, was developed more than 20 years ago, and together with its recent derivations (e.g. Microworlds®) supports learning of geometry, and in particular the sub-topic of location. LOGO® has useful applications across the primary school years. More recently developed programs for geometry are Cabri® and Geometer's Sketchpad®—commercial programs suitable for the upper years of primary schooling. The usefulness of these programs relates to the ways in which they enable students to engage in dynamic geometry activities (Battista, 2002).

Location and arrangement

The topic of location focuses on ways to characterise position within a space. In curriculum documents, this sub-strand has a range of labels—location, arrangement or position—and includes learning about coordinate systems. At more advanced levels, this includes developing notions associated with map reading—for example, coordinates and scales—and using a globe to find locations and learn about topics such as latitude and longitude.

● The language of location and movement

Locational or positional language is a key component of early location work. Students come to school with varying experiences of language relating to the notion of location. Some students will have a well-established location language whereas others may not. Teachers need to spend time developing this aspect of location. The language of location includes words to describe objects in relation to others—the chair is *behind* the table; the vase is *above* the cabinet. Other language can be related to movement—move *forward* one step; jump *backward* three squares.

Following is a list of terms associated with the language of location:

above	top
below	bottom
right	forward
left	backward
in front of	beside
behind	around
next to	

● Early location experiences

Students can begin to learn about coordinate systems in the early years (Dobler and Klein, 2002). Early experiences focus on students learning to read the grid references corresponding to a region where items appear. Typically, these involve an alpha-numerical pairing—that is, letters are used on one axis and numerals are used on the other. Each square on the map is characterised by a distinct alpha-numerical pair. When using a map such as this, students would work with letters on the x-axis (horizontal) and numerals on the y-axis (vertical). Using the alpha-numerical scale supports students' learning to read the x-axis first and y-axis second. This is important for both the topic of location and further work with graphs and functions. Students need to experience two aspects of location. First, the coordinates are given and the student identifies the site—what is located at B2? Second, the site is given and the student determines the coordinates—what are the coordinates for the bag of money?

As students develop knowledge of alpha-numerical scales, they can be introduced to the use of ordered numerical pairs. Thus activities involving reading the value on the x-axis, then reading the value on the y-axis, serve to develop knowledge of the more abstract notion of an ordered pair. The term 'ordered' is used in the sense that, by mathematical convention, the first value of the pair corresponds to the value of the x-axis, and the second corresponds to the value on the y-axis. Notation such as (3, 5) is used for an ordered pair.

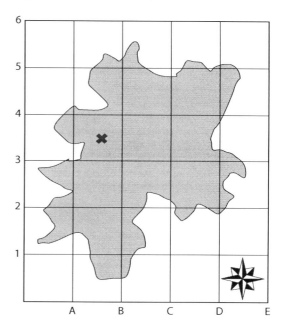

Simple map with alpha-numeric coordinates

● Later experiences of location

Once students have learnt to read ordered pairs, teaching can focus on associating each ordered pair with a point rather than a region. As part of the progression to associating each ordered pair with a point, each numeral on the axes corresponds with a grid line rather than an interval on the axis. This can be seen on the map above, where the numerals are aligned with the grid lines.

Initial learning experiences may involve the items being placed on intersecting grid lines. As students' fraction knowledge develops, items to be located can be placed at points where students can readily identify the point—this can involve simple fractions such as halves and quarters initially (1.5, 2–2.5), and later tenths (1.3, 1.8). This approach enables students to see the usefulness of finer gradations on the axes—points can be characterised more precisely.

Create problems for the students whereby they come to 'discover' the need for gradations in the pairings.

● Applications

Once students have developed knowledge of ordered pairs, they can engage in games such as Battleships, where the objective is to locate and sink your partner's ships (see 'Teaching idea—Battleships'). More realistic applications involve the use of maps to locate sites. Initially, this can involve local street directories, which typically use alpha-numerical scales and regions. Later experiences can focus on using maps in an atlas. An important point to keep in mind when working with maps in atlases is that coordinates are specified and read as follows—first latitude (North or South), then longitude (East or West). Thus the order of reading these map coordinates is at odds with the order of reading coordinates on a grid system—first horizontal (x-axis), then vertical (y-axis). Students may struggle with this at first, in part due to the order of reading the pairs, and also because of the use of four quadrants (NE, NW, SE, SW), but it is a life skill that is important for students to understand.

T E A C H I N G   I D E A

Battleships

Using grid paper, students put a number of ships on the paper. In pairs, students list off coordinates to identify and sink their partner's ships. The winner is the one who sinks all the ships first.

● Moving through space

A common teaching practice is for students to be provided with a map and then given a series of directions—such as move forward two units, turn right, then move forward x units. Such activities often use names of streets to incorporate a reality element to the task. Commonly, this activity may involve starting at a point, such as a railway station, and then navigating through streets, turns, distances, and so on to arrive at a particular place, which the student must name. This requires the coordination of a number of variables and is quite complex. It is a useful activity as it resembles out-of-school experiences of using maps to find sites or moving through new territories. Diezmann and Lowrie (2010) report that in this activity students often make errors by not coordinating the various elements of the task, such as overlooking one element—ordinal language (the third street) or positional language (at the school or into Lambert Rd) or positional language (turn right). They report that the development of this skill appears to plateau in the middle years around Grade 6 or 7.

● Location and technology

Research is beginning to emerge about students' map-reading skills. We have shown the typical teaching process, where elements are built into the curriculum progressively so that students learn the required mapping skills. This approach seems to limit what students can do when they use digital environments. Typically the experiences in school mathematics have been contained to two-dimensional space—such as maps on paper. However, in the ICT and games environments, young children are exploring 3D representations where objects are moved along the two dimensions of the screen but also appear to 'fall' into three-dimensional spaces as well. Lowrie (2003) has shown how young children (eight-year-olds) have been found to construct complex images of the maps beyond the immediate visual cues and to construct three-dimensional images. In his work in this area, Lowrie has shown how young children not only create three-dimensional mental maps of the worlds in which they play (such as Mario Brothers or Pokemon) but are capable of creating mental images of the maps outside the screen area. The impact of this has not yet formally been recognised in curriculum documents, but it challenges the static views of map reading that are currently presented. Furthermore, middle school students are often involved in complex computer games where they are required to navigate through worlds to solve problems. The challenges they confront in these worlds are not only useful in terms of problem-solving but also help to develop visualisation, as they create the mental maps of the worlds they will encounter as they move from one screen to the next. This information is not available to them but must be created before they are able to move successfully into the next screen. The impact of these games will have a profound effect on how map reading is to be conceptualised in the emerging theorising of curriculum and knowing. Clearly, these technologies will create new learning opportunites for spatial thinking and visualisation that were inconceivable in the traditional pencil-and-paper format of previous times.

A further advance in modern society is the increasing availability of Global Positioning Systems (GPS). Various tools are now available that allow students to undertake location activities. The GPS tools allow for activities such as orienteering or linking to other curriculum areas (such as science, to plot the location of species or to track animal migrations). While some of these tools may be expensive, others—tools such as car navigation tools—are more readily accessible and offer new representations of space and movement through space. Similarly, these tools are widely available on mobile phones and can readily be used to support learning activities. As mobile devices become increasingly available, access to such resources will enable new modes of teaching and learning to be made available to students and teachers.

● **Direction**

Direction is another aspect of the geometry topic of location, and involves learning about compass points. Initially, students learn about the four points north, south, east and west. Depending on syllabus requirements, this is extended to eight points in all by the inclusion of north-east, north-west, south-east and south-west, or to 16 points by also including NNE, ENE, NNW, WNW, SSE, ESE, SSW and WSW.

In students' initial learning of the four major points, north and south are usually learnt relatively easily, but some confusion can arise when learning east and west. Mnemonic devices—such as Never Eat Soggy Weetbix—can be used for learning these as the students rotate clockwise through the points. A common instructional activity involves students finding a particular point by following directions on a grid map—'two paces north, three paces west', and so on. Orienteering activities involving the use of a compass are also useful for this aspect of location.

The cultural relativity of direction is an interesting area of application. For most students growing up in Western societies, the concept of direction is relative—that is, it is relative to where you are located.

Terms such as left and right are commonly used for finding or giving directions. In some cultures, however, different applications are used. Some Indigenous Australian cultural groups use compass points in their everyday discussions (Harris, 1990). In other contexts, different methods are developed. For example, on Hawaii, relative (but significant) points are used as the terms of reference (Diamond Head, the mountains) so that compass points are not used when giving directions.

Teaching space in the middle years

The van Hiele levels of geometric thinking, summarised at the beginning of this chapter, are the foundation for instruction in the primary curriculum. Levels 4 and 5 would be the focus of activities in the primary years only informally. However, the middle years should be laying the foundation for students to achieve at level 4 in particular, where students develop the capacity to argue the case for why certain geometric situations are mathematically sound. This is the notion of proof, which is something that is not frequently used in the discourse of the middle years mathematics curriculum. However, calls have been made for proof to be an integral part of the middle years curriculum, in relation to providing convincing arguments in mathematics, and to show that 'what is offered as a convincing argument by one person must be accepted by others' (Harel and Sowder, 2007, p. 808). Harel and Sowder discuss the notion of a proof scheme, rather than proof in the formal sense, where members of a (classroom) community determine the parameters for ascertaining and persuading the community about the 'truth' of an assertion. In such a community, students would be given tasks where they must persuade others of the legitimacy of their thinking. Their argument is based on examples and generalisations rather than reference to an authoritative source, such as the teacher, the internet or the textbook. To emphasise the importance of engaging students in discussions about geometric relationships and notions, Harel and Sowder provide the following assessment item used to determine Year 11 students' understanding of mathematical argumentation and proof: 'Jim says, "If a 4-sided figure has all equal sides, it is a square." Which of the following figures might be used to prove that Jim is wrong?'

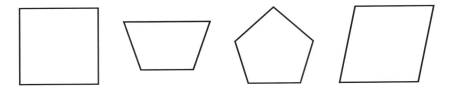

The number of students selecting each figure was 15, 31, 24 and 31 per cent respectively. The point is, middle years teachers need to move students beyond level 3 of the van Hiele model. Proof schemes require students to build mathematical arguments to convince others of their solutions, necessitating revisiting knowledge and understanding of properties of shapes and relationships.

To support students' movement between the van Hiele levels, extensive opportunities to explore and experience spatial activities are required. In the primary years, hands-on activities that include building, designing, constructing and representing are advocated for space. This must continue in the middle years, with tasks becoming increasingly sophisticated. For example, constructing shapes from 2 cm cubes, then drawing these designs on isometric paper, promotes visualisation and arrangement that supports reading and interpretation of the architectural drawings students will most likely encounter in their lives in the future.

■ REVIEW QUESTIONS

15.1 What common observations might you expect from your students when teaching basic geometric plane (2D) shapes that would confirm the van Hiele theory?

15.2 Explain the difference between nets and skeletal forms, using examples to illustrate the difference/s.

15.3 Describe three introductory activities that would be appropriate for the topic of tessellation.

15.4 Students need to have considerable experiences in language when working in the space strand. Give examples of early language that should be developed and of how you would go about providing learning experiences to develop this language.

15.5 Outline learning activities suitable for the topic of symmetry in the middle primary years.

15.6 In a multi-age classroom, there is considerable diversity in student learning. How would you use one stimulus map to cater for a diverse range of learning when teaching mapping skills to students?

15.7 Find out how other cultural groups use directions in their everyday lives. Compare and evaluate this with standard curriculum expectations.

Further reading

Battista, M. and Clements, D.H. (1998). Finding the number of cubes in rectangular cube buildings. *Teaching Children Mathematics*, 4(5), 258–64.

Britton, B. and Stump, S. (2001). Unexpected riches from a geoboard quadrilateral activity. *Mathematics Teaching in the Middle School*, 6(8), 490–3.

Clements, D. and Sarama, J. (2000). The earliest geometry. *Teaching Children Mathematics*, 7(2), 82–6.

Cockcroft, W.C. and Marshall, J. (1999). Educating Hannah: It's a what? *Teaching Children Mathematics*, 5(6), 326–9.

Fox, T. (2000). Implications of research on young children's understanding of geometry. *Teaching Children Mathematics*, 6(9), 572–6.

Koester, B. A. (2003). Prisms and pyramids: Constructing three-dimensional models to build understanding. *Teaching Children Mathematics*, 9(8), 436–42.

Liedtke, W. W. (1995). Developing spatial abilities in the early grades. *Teaching Children Mathematics*, 2(1), 12–18.

Lindquist, M.M. and Clements, D.H. (2001). Geometry must be vital. *Teaching Children Mathematics*, 7(7), 409–15.

Mitchelmore, M. (2000). Teaching angle measurement without turning. *Australian Primary Mathematics Classroom*, 5(2), 4–8.

Simpson, D. (2001). Shapes, shapes, everywhere! *Prime Number*, 16(3), 4–7.

Southwell, B. (1998). Give me space. *Reflections*, 23(1), 52–6.

Trafton, P.A. and Hartman, C. (1997). Exploring area with geoboards. *Teaching Children Mathematics*, 4(2), 72–5.

van Hiele, P.M. (1999). Developing geometric thinking through activities that begin with play. *Teaching Children Mathematics*, 5(6), 310–16.

——(2002). Similarities and differences between the theory of learning and teaching of Skemp and the van Hiele levels of thinking. In D. Tall and M. Thomas (eds), *Intelligence, learning and understanding in mathematics: A tribute to Richard Skemp* (pp. 27–47). Brisbane: Post Pressed.

Wheatley, G. (1990). Spatial sense and mathematics learning. *Arithmetic Teacher*, 37(6), 10–15.

White, P. and Mitchelmore, M. (1997). Sharpening up on angles. *Australian Primary Mathematics Classroom*, 3(1), 19–21.

Yackel, E. and Wheatley, G.H. (1990). Promoting visual imagery in young pupils. *Arithmetic Teacher*, 37(6), 52–8.

References

Ambrose, R.C. and Falkner, K. (2002). Developing spatial understanding through building polyhedrons. *Teaching Children Mathematics*, 8(8), 442–7.

Battista, M. (2002). Learning geometry in a dynamic computer environment. *Teaching Children Mathematics*, 8(6), 333–9.

Battista, M. and Clements, D.H. (1998). Finding the number of cubes in rectangular cube buildings. *Teaching Children Mathematics*, 4(5), 258–64.

Britton, B. and Stump, S. (2001). Unexpected riches from a geoboard quadrilateral activity. *Mathematics Teaching in the Middle School*, 6(8), 490–3.

Del Grande, J. (1990). Spatial sense. *Arithmetic Teacher*, 37(6), 14–20.

Diezmann, C. and Lowrie, T. (2010). Students as decoders of graphics in mathematics. In L. Sparrow, B. Kissane and C. Hurst (eds), *Shaping the future of mathematics education: Proceedings of the 33rd Annual Conference of the Mathematics Education Research Group of Australasia*. pp. 161–8. Fremantle: MERGA.

Dobler, C.P. and Klein, J.M. (2002). First graders, flies, and a Frenchman's fascination: Introducing the Cartesian coordinate system. *Teaching Children Mathematics*, 8(9), 540–5.

Fox, T. (2000). Implications of research on young children's understanding of geometry. *Teaching Children Mathematics*, 6(9), 572–6.

Harel, G. and Sowder, J. (2007). Towards comprehensive perspectives on the learning and teaching of proof. In F. Lester (ed.), *Second handbook of research on mathematics teaching and learning* (pp. 805–42). Charlotte, NC: Information Age.

Harris, P. (1990). *Mathematics in a cultural context: Aboriginal perspectives on space, time and money.* Geelong: Deakin University Press.

Lowrie, T. (2003). Posing problems in ICT-based contexts. In L. Bragg, C. Campbell, G. Herbert and J. Mousley (eds), *Mathematics education research: Innovation, networking and opportunity. Proceedings of the 26th Annual Conference of the Mathematics Education Research Group of Australasia* (Vol. 1, pp. 499–506). Geelong: MERGA.

Lowrie, T. and Diezmann, C. (2007). Solving graphics problems: Student performance in junior grades. *Journal of Educational Research*, 100(6), 369–78.

——(2009). Solving graphics tasks: Gender differences in middle-school students. *Learning and Instruction. Online version available at <www.sciencedirect.com/science/article/B6VFW-4Y0C1DY-1/2/20578a55dc698a7ee395a5e8dc3fe0fb>.*

Mitchelmore, M. (2000). Teaching angle measurement without turning. *Australian Primary Mathematics Classroom*, 5(2), 4–8.

Morrow, L.J. (1991). Geometry through the standards. *Arithmetic Teacher*, 38(8), 21–5.

Owens, K., Reddacliff, C., Gould, P. and McPhail, D. (2001). Changing the teaching of space mathematics. In J. Bobis, B. Perry and M. Mitchelmore (eds), *Proceedings of the 25th Annual Conference of the Mathematics Education Research Group of Australasia* (Vol. 2, pp. 402–9). Sydney: MERGA.

Paznokas, L. (2003). Teaching mathematics through cultural quilting. *Teaching Children Mathematics*, 9(5), 250–6.

Pegg, J. (1985). How children learn geometry: The van Hiele theory. *The Australian Mathematics Teacher*, 14(2), 5–8.

Pegg, J. and Davey, G. (1989). Clarifying level descriptors for children's understanding of some basic 2-D geometric shapes. *Mathematics Education Research Journal*, 1(1), 16–27.

Southwell, B. (1998). Give me space. *Reflections*, 23(1), 52–6.

Trafton, P.A. and Hartman, C. (1997). Exploring area with geoboards. *Teaching Children Mathematics*, 4(2), 72–5.

van Hiele, P.M. (1999). Developing geometric thinking through activities that begin with play. *Teaching Children Mathematics*, 5(6), 310–16.

Whitaker, R. (2001). Tessellating can be fun. *Vinculum*, 38(2), 10–12.

Index

nominal numbers 143–4

Number Framework (New Zealand) 151–2, 217; counting stages 151–2; nine global strategies 151, 217; part-whole stage 151

number sense 13, 130–2; developing 162–3

number: algebra and 267; benchmarks 141–2; classification 132–3; concrete representation 130; early 130–58; intuitive understanding 130; knowledge at school entry 131–2; line 217; linguistic representation 130; patterns 132; strand of curriculum 13–14, 16; study 13; symbolic representation 130; *see also* composite numbers, geometric numbers, palindromic numbers, pre-number, prime numbers

numbers: grouping 148–9; symbolizing 148; types 143–5; *see also* cardinal numbers, ordinal numbers, nominal numbers, numerals

numeracy 19; political aspect 19; theory and 23

numerals: sequencing 145; as symbolic numbers 144–5; writing 144–5; *see also* numbers

numeration 145–54; *see also* base 10 numeration system, Hindu-Arabic system

observation 95–8; records of 96–7; schedule 97; performance tasks 98

one-hundred board 176–8

open-ended tasks 56–7

operations, study of 13–14; on whole numbers, *see* whole numbers

ordinal numbers 143

packing 344

palindromic numbers 270–1

patterns 8–9, 132, 258, 259–66; activities 133; copying 260; describing 260; early activities 260–1; extending 260; growing 259, 261–2; language of 259; making 260; number 259, 262–3; odd and even numbers 263–4; pre-number 133; as problem-solving strategy 118; relationships 264; repeating 259, 260–1; *see also* function machines

pedagogy 6–8, 12; in diverse classroom 62; explicit 45–6; middle years 18; productive pedagogies 6

peer assessment 102

pen-and-paper: practice 161; testing 92

per cent 247–54; building conceptual understanding 247, 248–9; increases and decreases 253–54; increases of 100 per cent 252–3; increases of 200 per cent 252; increases of over 100 per cent 251; language 249–50; mental computation 250–1; multiple meanings 247–8; part and complement 250–1; real-world applications 247; relationship with decimals 247, 248; perimeter, area and 301–2

physically disabled students 53

Piaget, Jean 15, 24–5, 137, 277; conservation of length 277; stage theory 24; *see also* conservation

place value 135, 145–54, 202, 210; decimals 231–4; developing 146–8; model 162; two-digit numbers 146

planning 65–81; content 67–72; cyclical approach to, 88–9; focus 68–72; individual 67; levels 66–7; trans-disciplinary 72; long-term 66; medium term 66; short-term 67; principles 67; reasons for 67–8; school-wide 67; for substantive learning 81; team 67

Platonist theory 11

19996277R00221

Printed in Great Britain
by Amazon